RAPID REVISION NOTES

NOTES

A LEVEL

PHYSICS

Mechanics and
Properties of Matter

RAPID REVISION NOTES
A LEVEL
PHYSICS
Mechanics and
Properties of Matter

BY C BOYLE
B.Sc. (Hons), P.G.C.E., M.Inst.P., Grad. I.M.A.
Lecturer in Physics and Mathematics,
North Cheshire College,
Warrington

General Editor – Professor A. J. B. Robertson,
Professor of Chemistry, King's College,
London

CELTIC
REVISION AIDS

Celtic Revision Aids
30–32, Gray's Inn Road,
London WC1X 8JL.

© C.E.S.

First Published 1982

ISBN 0 86305 118 9

Composition by
Filmtype Services Limited,
Scarborough, North Yorkshire.

Printed and bound in Great Britain by
Cox & Wyman Ltd, Reading

General Editor's foreword

The Rapid Revision 'A' Level Series is designed for students preparing for G.C.E. 'A' Level, Scottish Highers, Intermediate University and similar examinations. They are comprehensive and may be used either on their own for revision or as a complement to set books and text books. The notes are organised to help students remember facts and to pin-point areas of difficulty. Practice questions are included at the end of each section with answers to the numerical problems at the end of the text. Where necessary, clearly worked through examples have been included in the text.

I am sure students will find these books very helpful.

A. J. B. Robertson

M.A., Ph.D., D.Sc., C.Chem., F.R.S.C.

Professor of Chemistry,
King's College London,
University of London

Formerly Fellow of St John's College, Cambridge

Author's foreword

The aim of this book is to give the prospective 'A' level physics examinee a concise but complete revision course in this particular branch of the subject.

The material is grouped according to the principal topics and concepts included in the syllabuses of the major examining boards.

After each concept is introduced, specimen questions are worked through as examples and then problems are provided for the student to attempt for himself. Finally, at the end of the book, omnibus examples and questions are provided. This is felt to be necessary since examination questions usually involve simultaneous applications of concepts and equations not related to each other.

Where necessary, answers have been rounded to four significant figures in order that students without a calculator need not feel at a disadvantage. Also, in most problems, g has been taken to be $10\,\text{ms}^{-2}$, although not always. The aim of this is twofold: firstly to ease calculation and secondly to persuade the student that g is not an absolute constant!

In labelling graphical axes the convention of Quantity (Unit) has been adopted as suggested by The Institute of Physics.

In writing this book, the author has used his experience as a lecturer and examiner. Common faults amongst examinees are answering questions not asked and offering vague definitions. If the student learns the definitions presented and successfully attempts the problems, he should be able to approach the examinations with confidence.

Christopher Boyle

Contents

Contents

1 UNITS AND DIMENSIONS

The exact description of a physical quantity consists of the product of two factors: a pure number and a given unit.

The three fundamental units are: length, measured in metres and denoted by [L]; mass, measured in kilograms and denoted by [M] and time, measured in seconds, denoted by [T].

The dimensions of a physical quantity is its relationship to [L], [M] and [T].

Quantity	Dimensions
Distance	$[L]$
Area	$[L^2]$
Volume	$[L^3]$
Velocity	$[LT^{-1}]$
Acceleration	$[LT^{-2}]$
Force	$[MLT^{-2}]$
Pressure	$[ML^{-1}T^{-2}]$
Work and Energy	$[ML^2T^{-2}]$

The dimensions of any physical quantity may be derived by the use of dimensions in any defining equation.

Examples

1 Derive the dimensions of density

By definition, density $= \dfrac{\text{mass}}{\text{volume}} = \dfrac{[M]}{[L^3]} = [ML^{-3}]$

2 Derive the dimensions of power

By definition, power = rate of doing work $= \dfrac{dW}{dt} = \dfrac{[ML^2T^{-2}]}{[T]}$

$$= [ML^2T^{-3}]$$

(Note: the dimensions of change in any quantity are the same as those of the quantity itself.)

There are two uses for dimensions: firstly, to check a formula and secondly to derive an equation. In checking a formula we use the fact that in any equations the dimensions on both sides must be identical if the formula is correct. In deriving an equation we bear in mind that the method only enables us to find a proportional relationship between variables. Constants of proportionality cannot be found by the method of dimensions.

Examples

1   To check the 'pressure-head' formula

$$p = \rho gh$$

p = pressure, ρ = density of fluid, h = height of pressure head, g = acceleration due to gravity.

Dimensions of left hand side of equation: $[ML^{-1}T^{-2}]$ (see table)

Dimensions of right hand side of equation: $[ML^{-3}] \times [LT^{-2}] \times [L]$

$$= [ML^{-1}T^{-2}]$$

Dimensions of both sides are identical therefore the equation is correct.

2   To derive an equation for the period of oscillation, T, of a simple pendulum.

We assume that the amplitude of oscillation is small and ignore all dissipative forces.

Suppose the period depends upon the mass of the pendulum bob, the length of the pendulum and the acceleration due to gravity. Since the exact nature of the dependence is not known, we assume arbitrary powers for each variable:

Period $\propto$ (Mass)^x × (Length)^y × (Acc. due to gravity)^z

This is converted to an equation by introducing a constant of variation **K**.

$\therefore$ Period = K × (Mass)^x + (Length)^y × (Acc. due to gravity)^z

Change each side of the equation into the dimensional equivalent:

Dimensions of Period = $[T]$

Dimension of K = **none: a constant has no dimensions**

Dimensions of Mass = $[M]$

Dimensions of Length = $[L]$

Dimensions of Acceleration due to gravity = $[LT^{-2}]$

$$\therefore   [T] = [M]^x [L]^y [LT^{-2}]^z$$

Collecting together powers on the right hand side of the equation

$$[T] = [M]^x [L]^{y+z} [T]^{-2z}$$

We now use the mathematical fact that powers on each side of an equation should be identical, remembering that any quantity raised to the zero[th] power is unity.

For the left hand side of the equation:

Power to which $[T]$ is raised $= 1$ \qquad (a)

Power to which $[M]$ is raised $= 0$ \qquad (b)

Power to which $[L]$ is raised $= 0$ \qquad (c)

For the right hand side of the equation:

Power to which $[T]$ is raised $= -2z$ \qquad (d)

Power to which $[M]$ is raised $= x$ \qquad (e)

Power to which $[L]$ is raised $= y + z$ \qquad (f)

From (b) and (e) $x = 0$

From (a) and (d) $1 = -2z$ or $z = -\frac{1}{2}$

From (c) and (f) $0 = y + z$ or $0 = y - \frac{1}{2}$ (substituting value of z)

$$\therefore y = \frac{1}{2}$$

Using these values in the original equation:

Period = K (mass)$^{\circ}$ (Length)$^{\frac{1}{2}}$ (Acc. due to gravity)$^{-\frac{1}{2}}$

or

$$\text{Period} = K \sqrt{\frac{\text{Length}}{\text{Acc. Grav}}}$$

By comparing the equation with the well-known formula:

$$T = 2\pi \sqrt{\frac{l}{g}}$$

we see that the value of K is 2π. Note that a rigorous derivation is necessary to conclude this fact. Also, note that any false assumptions – e.g. a dependence on mass – disappears 'in the mathematics'.

Practice questions

1 Derive the dimensions of the moment of a force.
2 Derive the dimensions of relative density and comment upon your result.

Problems

3 Use the method of dimensions to check the formula for centripetal acceleration

$$a = \frac{v^2}{r}$$

a = acceleration, v = speed of body, r = radius or orbit.
4 Use the method of dimensions to derive an equation for velocity of deep water waves (Hint: suppose it to depend upon the density of water, the acceleration due to gravity and the wavelength).

2 ELEMENTARY QUANTITIES IN DYNAMICS

Mass
The mass of a body is defined as the quantity of matter contained in the body. **Units: kilogram, Dim: M.** The mass of an object is invariant i.e. it does not vary from place to place.

Weight
The weight of a body is defined as the force acting on it due to the gravitation or attraction of the planet on which it is situated. Naturally, for all practical purposes, the planet is earth! **Unit: Newton, Dim: MLT^{-2}.** The weight of an object is not constant but depends upon location.
The relationship between mass and weight will be considered later.

Displacement
Displacement is defined as distance moved in a specified direction.

Unit: metres with direction specified
Dim: L

Speed
Speed is defined as rate of change of distance moved with time.

$$\textbf{Speed} = \frac{\textbf{dx}}{\textbf{dt}}$$

x = distance   t = time
Unit: ms^{-1} dim: LT^{-1}

Velocity
Velocity is defined as rate of change of distance moved in a specified direction (i.e. displacement) with time.

$$\textbf{Velocity } v = \frac{\textbf{dx}}{\textbf{dt}}$$

x = displacement   t = time
Unit m s^{-1} in given direction
dim: LT^{-1}

Note: For simple cases of straight line motion, the terms speed and velocity are synonymous each being numerically identical. Thus a speed of 6 m s^{-1} moving in a straight line has a velocity of 6 m s^{-1} **in the direction of the straight line.**

However, the situation is more complex in the case of circular motion – see later.

Acceleration
Acceleration is defined as rate of change of velocity moved with time.

$$\text{Acceleration } a = \frac{dv}{dt}$$

Unit: $m\,s^{-2}$
dim: LT^{-2}

Alternative representations:
1 Since

$$v = dx/dt$$

then

$$a = \frac{dv}{dt} = \frac{d^2x}{dt^2}$$

2 Using 'Chain-Rule' of calculus

$$a = \frac{dv}{dt} = \frac{dv}{dx} \times \frac{dx}{dt} = \frac{dv}{dx} \times v$$

A negative acceleration corresponds to a deceleration or retardation.

Examples
1 A body travelling at a constant speed travels 88 km in 1 hr. Calculate its speed, in metres sec^{-1}.
Since the speed is constant then:

$$\text{Speed, in } m\,s^{-1} = \frac{\text{distance in m}}{\text{time in s}} = \frac{88 \times 1000}{3600} = \frac{220}{9}\ m\,s^{-1}$$

2 A body moving with an initial velocity of 50 m s^{-1} uniformly changes its velocity to 25 m s^{-1} in 5 s. Find its acceleration.
Since the acceleration is uniform:

$$\text{Acceleration} = \frac{\text{Velocity change}}{\text{Time taken}} = \frac{25 - 50}{5} = -5\ m\,s^{-2}$$

Minus sign signifies a retardation.

Rectilinear motion

Equations of uniformly accelerated motion

If an object accelerates at a m s^{-2} from an initial velocity of u m s^{-1} to a final velocity of v m s^{-1} in a time t s whilst travelling a distance in a given direction s m then:

$$s = ut + \tfrac{1}{2}at^2$$
$$v = u + at$$
$$v^2 = u^2 + 2as$$
$$s = \left(\frac{u + v}{2}\right)t$$

$a = \frac{v-u}{t}$

Notes

In the case of deceleration replace a by −a. For bodies falling under gravity, replace a by g.

Graphical representation

From a graph of velocity against time, we can obtain the acceleration – by finding the gradient of the graph – and the total distance travelled – by finding the area under the graph. This is illustrated in Fig. 1.

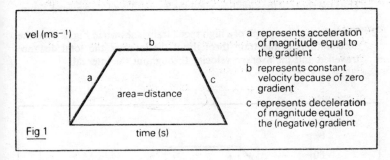

| vel (ms^{-1}) | |
| area=distance | |

a represents acceleration of magnitude equal to the gradient

b represents constant velocity because of zero gradient

c represents deceleration of magnitude equal to the (negative) gradient

Fig 1   time (s)

Examples

1   A body is projected vertically upwards with a velocity of 20 m s^{-1}. Find the maximum height and the time elapsing before it is at a height of 12 m. (Assume g = 10 m s^{-2}.)

   We note that the motion is symmetrical: the time of upward motion is identical to that of the downward motion.

   Often problems of this type can be solved in a variety of ways – the technique being to use a minimum number of equations to ellicit the maximum information.

In order to find the maximum height use:

$$v^2 - u^2 + 2as$$

$u = 20 \, m \, s^{-1}$

$v = 0 \, m \, s^{-1}$ (at maximum height the object is at rest)

$a = -g = -10 \, m \, s^{-2}$ (acting to cause deceleration)

$0 = 400 - 20 \, s$

$s = 20 \, m$

The object will be at a height of 12 m both on its upward and downward journeys.

For the time use

$$s = ut + \tfrac{1}{2} at^2$$
$$12 = 20 \, t - 5 \, t^2$$

Solving this by formula yields

$$t = 0{\cdot}735 \, s \quad \text{or} \quad 3{\cdot}265 \, s$$

Thus the roots of the equation yield the 'upward' and 'downward' times.

2   The velocity time graph of a high speed train is shown in Fig. 2. Find the accelerations of the train throughout the journey, the total distance travelled and the average velocity throughout the interval.

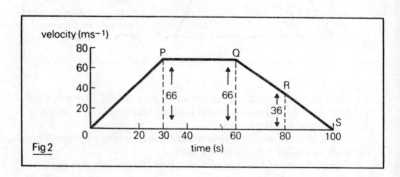

Fig 2

For OP, the constant acceleration is the gradient $= \dfrac{66}{30} = 2 \cdot 2 \, \text{m s}^{-2}$

For PQ, the acceleration is $0 \, \text{m s}^{-2}$

For QR, the acceleration is $-\left(\dfrac{66 - 36}{20}\right) = -1 \cdot 5 \, \text{m s}^{-2}$

For RS, the acceleration is $-\dfrac{36}{20} = -1 \cdot 8 \, \text{m s}^{-2}$

Total distance

$\quad$ = Area under graph

$\quad = (\tfrac{1}{2} \times 30 \times 66) + (66 \times 30) + (\tfrac{1}{2} \times 36 \times 20) + \left(20 \times \dfrac{66 + 36}{2}\right)$

$\quad = 4350 \, \text{m}$

Average velocity $= \dfrac{\text{Total distance}}{\text{Total time}} = \dfrac{4350}{100} = 43 \cdot 5 \, \text{m s}^{-1}$

Practice questions

1 Find the constant speed, in m s^{-1}, of an object which travels $7 \cdot 2$ km in 1 hr.
2 Find the constant acceleration of an object which accelerates from a velocity of $10 \, \text{m s}^{-1}$ to $70 \, \text{m s}^{-1}$ in 15 s.
3 If a sprinter can start with a velocity of $8 \, \text{m s}^{-1}$ and run with uniform acceleration find, graphically, the greatest velocity reached in running the 100 metres in 10 sec and the necessary acceleration.
4 A body is projected vertically upwards with a velocity of $30 \, \text{m s}^{-1}$. Find how long it takes to reach its highest point and the distance it ascends in the third second of its motion. (Assume $g = 10 \, \text{m s}^{-2}$.)

3 SCALARS AND VECTORS

A scalar quantity is one which has magnitude only.

Examples: Distance, speed, volume, temperature, mass, energy and power.
Scalars are added arithmetically.

A vector quantity is one which has both magnitude and direction.

Examples: Displacement, velocity, acceleration, momentum and force (including weight!).

Vectors are added algebraically and can be completely represented by straight lines, drawn to scale, in the appropriate direction.

Combination of vectors

The resultant of two or more vectors is the single vector to which they are collectively equivalent. The process of finding the resultant is called **compounding** or **composition**. The same rule holds for all vectors, so we shall consider the self-evident case of displacements.

Rules for the composition of two vectors

(We shall not discuss the trivial case of vectors which **are parallel or anti-parallel**, when they are **added** or **subtracted**.)

1   Triangle rule

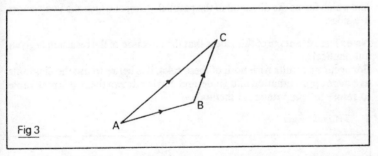

Fig 3

Consider a displacement **AB** (i.e. from A to B), followed by a displacement BC. The resultant displacement is given by **AC**. Notice that we place the second vector at the end of the first and the direction of the arrow on the resultant is such as to oppose the direction of the arrows on the initial vectors (see Fig. 3).

Triangle Rule: The resultant of two vectors, represented in magnitude and direction by straight lines placed end to end, is given by the line which completes the triangle in a direction such as to oppose cyclic travel around the sides.

2   Parallelogram Rule

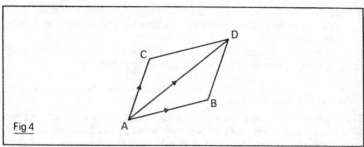

Fig 4

Consider a displacement **AB** to be compounded with a displacement **AC**. The resultant displacement is given by **AD**. Notice that both vectors start at a common origin and the direction of the arrow on the resultant is such as to point away from the common origin. (See Fig. 4.)

Parallelogram Rule: The resultant of two vectors represented in magnitude and direction by straight lines, drawn from a common origin, is given by the line which forms the diagonal of the parallelogram of which the two vectors are sides.

Note: The advantage of this rule is that the direction of the resultant is given automatically.
For accurate results with both of these rules, it is better to use the diagrams as a mere approximation and guide, and, having drawn them, it is preferable to resort to computational methods.

1   Triangle Rule

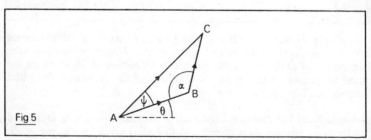

Fig 5

Magnitude of resultant found by invoking cosine rule:

$$AC^2 = AB^2 + BC^2 + 2AB\,BC\,Cos\,\alpha$$

Direction found by invoking same rule:

$$Sin\,\psi = \frac{BC}{AC}\,Sin\,\alpha$$

$$\therefore\ \psi = Sin^{-1}\left(\frac{BC}{AC}\,Sin\,\alpha\right)^{\circ}$$

and if AB is inclined at θ° to the horizontal then the inclination of the resultant to the horizontal is given by

$$\psi^{\circ} + \theta^{\circ} = \left(Sin^{-1}\left(\frac{BC}{AC}\,Sin\,\alpha\right) + \theta\right)^{\circ}$$

2 Parallelogram Rule

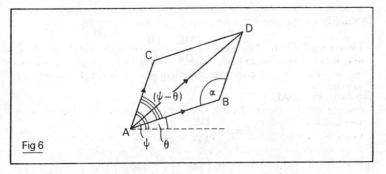

Fig 6

Firstly we must use some elementary geometrical facts: If **AB** is at θ° to the horizontal whilst **AC** is inclined at ψ° to the horizontal, then the angle between the vectors is $(\psi - \theta)^{\circ}$.

$$\text{Thus } \alpha = (180 - (\psi - \theta))^{\circ}$$

As before:

$$AD^2 = AB^2 + BD^2 + 2AB\,BD\,Cos\,\alpha$$

In order to find the inclination of the resultant we consider a modified version of Fig. 6:

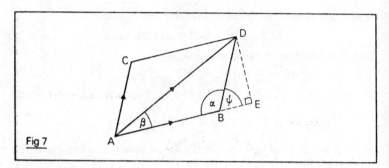

Fig 7

We wish to find β.
AB is produced to **E** and **DE** is drawn perpendicular to **BE**.
By geometry, angle DBE is ψ.

Now by definition

In △DBE

$$\text{Sin}\,\psi = \frac{DE}{DB} = \frac{DE}{AC}$$

$$\therefore DE = AC\,\text{Sin}\,\psi$$

Similarly, in **△DAE**

$$\text{Sin}\,\beta = \frac{DE}{AD} = \frac{AC}{AD}\,\text{Sin}\,\psi$$

$$\therefore \quad = \left(\text{Sin}^{-1}\left(\frac{AC}{AD}\,\text{Sin}\,\psi\right)\right)^{\circ}$$

Note: If the original vectors are perpendicular to each other then the triangle becomes a right-angled triangle and the parallelogram becomes a rhombus and the associated mathematics becomes correspondingly easier.

Rules for the composition of more than two vectors

1 One can group the vectors in pairs and successively apply either the triangle or parallelogram rule.
2 One can approach the problem diagrammatically by placing the vectors end to end, the resultant being given by that line which closes the polygon so formed and its direction being from the start of the first vector to the end of the last vector, as in Fig. 8.

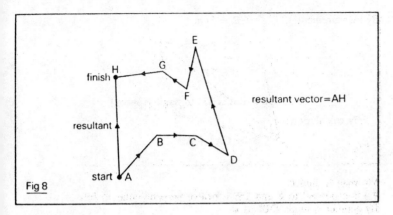

Fig 8

Resultant Vector = AH

Examples

1 A boat travelling at $3\,m\,s^{-1}$ sets course across a stream flowing at $4\,m\,s^{-1}$. Find the resultant velocity of the boat and its direction. Since velocity is a vector, we solve this problem using the above rules.

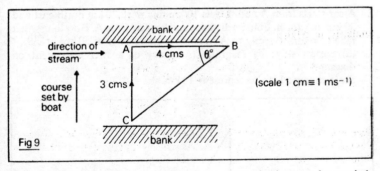

Fig 9

The effect of the flowing stream is to cause the boat to be carried downstream whilst it is crossing.

Let **AB** = velocity of stream

Let **CA** = course of boat

Then   resultant = **CB**

By measurement $\qquad$ CB = 5 cm

$$= 5\,m\,s^{-1}$$

$$\theta = 37°$$

$\therefore$ Direction is 37° to bank (approx).

By calculation:

$$CB^2 = CA^2 + AB^2$$

$$= 3^2 + 4^2$$

$$= 9 + 16 = 25$$

$$\therefore CB = 5\,m\,s^{-1}$$

$$Tan\,\theta = \tfrac{3}{4}$$

$$\therefore \theta = 36°\ 52'$$

Direction is 36° 52' to bank

2 A car starts from A and travels 10 km due west, 20 km north-east and 30 km due north. Find its distance and bearing from A. How long will it take to return directly to A at an average velocity of 50 km hr^{-1}?
(a) Easiest way: **by diagram**. (Accuracy of answer dependent on diagram)
Scale chosen: 1 cm to represent 10 km.

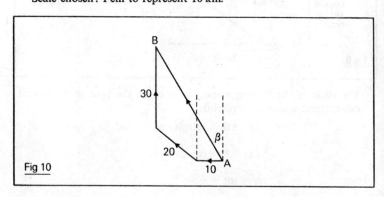

Fig 10

AB measured to be 5 cm.

∴ **distance from A = 50 km**

Bearing, angle β, measured to be 29°.

∴ **bearing N 29° W**

$$\text{Time} = \frac{\text{Distance}}{\text{Velocity}} = \frac{50}{50} = \textbf{1 Hr}$$

(b) Rigorous way: **by successive combination of vectors.**

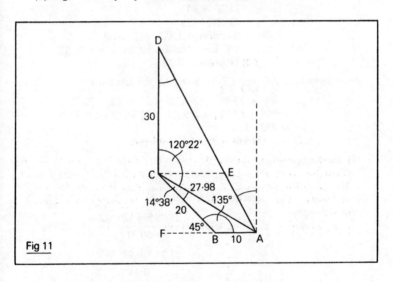

Fig 11

(i) Compound AB and BC to give AC.

$$AC^2 = 10^2 + 20^2 - 2 \times 10 \times 20 \, \text{Cos} \, 135$$
$$= 10^2 + 20^2 + 2 \times 10 \times 20 \, \text{Cos} \, 45$$
$$= 782 \cdot 8$$
$$\therefore \; AC = \textbf{27·98 km}$$

(ii) Find angle BCA

$$\frac{10}{\text{Sin} \angle \text{BCA}} = \frac{27 \cdot 98}{\text{Sin} \, 135°}$$

$$\therefore \ \text{Sin} \angle \text{BCA} = \frac{10 \sin 135°}{27 \cdot 98}$$

$$= \frac{10 \sin 45°}{27 \cdot 98}$$

$$= 0 \cdot 2527$$

$$\therefore \ \angle \text{BCA} = 14° \ 38'$$

(iii) Find angle DCA

$$\angle \text{BCE} = \angle \text{CBF} = 45°$$
$$\angle \text{DCE} = 90°$$
$$\therefore \ \angle \text{DCB} = 135°$$
$$\text{But } \angle \text{DCB} = \angle \text{DCA} + \angle \text{BCA}$$
$$\therefore \ \angle \text{DCA} = \angle \text{DCB} - \angle \text{BCA} = 120° \ 22'$$
$$\therefore \ \angle \text{DCA} = 120° \ 22'$$

(iv) Compound AC and CD to give AD, the resultant:

$$AD^2 = 30^2 + 27 \cdot 98^2 - 2 \times 30 \times 27 \cdot 98 \cos 120° \ 22'$$
$$= 30^2 + 27 \cdot 98^2 + 2 \times 30 \times 27 \cdot 98 \cos 59° \ 38'$$
$$= 2531 \cdot 4$$
$$\therefore \ AD = 50 \cdot 31 \ \text{km}$$

(v) Find angle CDA

$$\frac{27 \cdot 98}{\sin \angle \text{CDA}} = \frac{50 \cdot 31}{\sin 120° \ 22'}$$

$$\therefore \ \text{Sin} \angle \text{CDA} = \frac{27 \cdot 98 \sin 120° \ 22'}{50 \cdot 31}$$

$$= \frac{27 \cdot 98 \sin 59° \ 38'}{50 \cdot 31}$$

$$= 0 \cdot 4799$$

$$\therefore \ \angle \text{CDA} = 28° \ 41'$$

But this is the 'size' of the bearing by elementary geometry.

$$\therefore \ \textbf{Bearing} = \textbf{N28° 41'W}$$
$$\therefore \ \textbf{Distance \& bearing from A} = \textbf{50·31 km, N28° 41'W}$$
As before

$$\text{Time} + \frac{\text{Distance}}{\text{Velocity}} = \frac{50 \cdot 31}{50} = 1 \cdot 006 \ \text{hr}$$

Resolution of vectors

The reverse process of composition is referred to as resolution. In general the vector is resolved into two other vectors or components using the triangle or parallelogram rule in reverse and the most common case is when the components are perpendicular.

Example

A mass of 5 kg rests on a place inclined at 30° to the horizontal. Find the components of its weight both parallel and perpendicular to the place. (Assume $g = 10 \text{ m s}^{-2}$.)

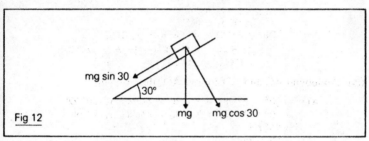

Fig 12

The weight acts vertically downwards (always!). Components of weight parallel to plane

$$= mg \sin 30° - 5 \times 10 \times 0.5 = 25 \text{ N}$$

Components of weight perpendicular to place

$$= mg \cos 30° = 5 \times 10 \times 0.866 = 43.3 \text{ N}$$

Practice questions

1 A boy walks from a place 0 a distance 6 km due east and then 8 km due south. Show that he is 10 km from 0 in a direction $\tan^{-1}(\frac{4}{3})$ south of east. How far due west must he now walk in order to reach a position south west of 0?

2 Vectors of magnitude 4, 10 and 6 units are acting in NE, E and SE directions respectively. Find the magnitude and direction of the resultant.

3 An object is moving in a straight line with a velocity of 10 m s^{-1}. Find the component of its velocity in a direction inclined at an angle of 30° to its direction of motion when the other component makes an angle of (1) 60° and (11) 90° with the direction of motion.

4 FORCE

Force is defined as that which changes, or tends to change, a body's state of rest or uniform motion in a straight line. **Unit: Newton, Dim: MLT^{-2}**.

Types of force

1 Weight
Weight is defined as the force acting on a body due to the earth's gravitational field.

It is directed vertically downwards, or, more correctly to the centre of the earth.

2 Upthrust
When an object is partially or totally immersed in a fluid, it experiences a force in a direction opposing its weight. This force is known as upthrust.

The magnitude of the upthrust is given by Archimedes' Principle: When an object is partially or totally immersed in a fluid, it experiences an upthrust equal to the weight of fluid displaced.

Proof (Applies to all fluids – liquids and gases.)

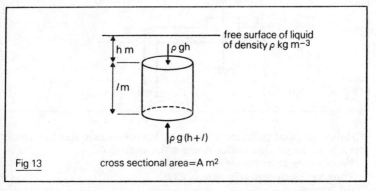

Fig 13   cross sectional area = A m²

Consider a cylinder of length l m, area of cross-section A m², immersed with its axes vertical and its upper face at a depth h m below the free surface of the liquid of density ρ kg m⁻³.

The downward pressure on the upper face = $\rho g h$ Pascals

∴ Total downward force on this   = $\rho g h A$ Newtons

The upward pressure on the lower face = $\rho g(h + l)$ Pa

∴ Total upward force on this   = $\rho g(h + l) A$ N

Resultant upward force or upthrust $= \rho g(h + l)A - \rho ghA$
$\qquad\qquad\qquad\qquad\qquad\qquad\qquad = \rho glA$ N
Now volume of cylinder $\qquad\qquad = Al$ m^3

$\qquad\qquad\therefore$ volume of liquid displaced $= Al$ m^3
$\qquad\qquad\therefore$ weight of liquid displaced $= Al\rho g$ N
$\qquad\qquad\therefore$ **Upthrust = Weight of Liquid Displaced**

Note that if the weight of the body exceeds the upthrust (i.e. if the density of the object exceeds that of the fluid), the body sinks.

3 Friction

Whenever two surfaces move (or attempt to move) with respect to each other, a force arises to oppose that motion. This force is known as friction. **Friction is that force which opposes motion.**

(a) Coefficient of Static Friction

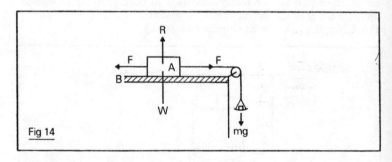

Fig 14

To investigate the Coefficient of Static Friction between surfaces A and B we use a block to provide surface A and a flat surface, B.

Force acting to accelerate **A** $\quad = mg$ N
Force acting to oppose motion $= F$ N
(i.e. friction)
Weight of block $\qquad\qquad\qquad = W$ N

Normal Reaction (reaction force normal to surfaces) $= R = W$ N. Masses added to increase mg until the block just begins to move. The frictional force now acting is known as the limiting frictional force, F N. When this happens $F = mg$.

Experiment is repeated with masses on the block so as to increase W and hence R. New values for F found.

We find the ratio:

$$\frac{\text{Limiting frictional force}}{\text{Normal Reaction}} = \frac{F}{R} = \text{Constant}$$

Constant known as Coefficient of Static Friction, μ_s

$$\therefore \ \mu_s = \frac{F}{R} = \frac{mg}{W}$$

(b) Coefficient of Dynamic Friction

In contrast to the case of static friction, when one of the objects is just on the point of slipping, dynamic friction occurs when the object moves with a uniform velocity.

The experiment is repeated but now F is the dynamic friction force which acts when the object is moving with a constant velocity.

We define the ratio:

$$\frac{\text{Dynamic Frictional Force}}{\text{Normal Reaction}} = \frac{F}{R} = \text{Constant}$$

Constant known as Coefficient of Dynamic Friction, μ_D

$$\therefore \ \mu_D = \frac{F}{R} = \frac{mg}{W}$$

Note: (1) $\mu_s > \mu_D$
        (2) Frictional force independent of area of contact.
An alternative method of finding μ_s involving no weights is shown in Fig. 15.

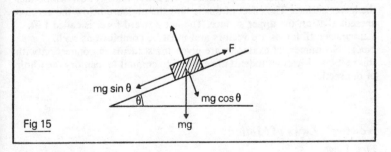

Fig 15

Plane tilted until slipping occurs
Resolving the forces we have

$$\mu_s = \frac{F}{R} = \frac{mg \, Sin \, \theta}{mg \, Cos \, \theta} = Tan \, \theta$$

$$\mu_S = \mathbf{Tan} \, \theta$$

4   Tension

When a force is applied to an object which transmits the force throughout its bulk we refer to the generation of a tension in the object. The direction of the tension is that of the applied force.

Examples: masses suspended on a string, pulling an object with a rope, holding up floors etc. with vertical columns.

5   Lift

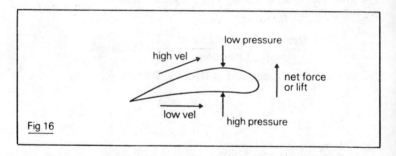

Fig 16

Fig. 16 shows an aerofoil i.e. an aircraft wing. The shape leads to a high velocity flow of air over the top surface and a lower velocity flow of air under the lower surface. The effect of this is that the pressure below the aerofoil exceeds that on the upper surface. The net upward force is called **Lift**.

Remember: all forces are vectors and must be combined as such.

Note: No numerical examples are given for students in connection with this section. However students are advised to commit to memory the whole of the section.

Newton's Laws of Motion

First Law

Every body continues in a state of rest or uniform motion in a straight line unless compelled to do otherwise by an external force.

Second Law
The rate of change of momentum (mass × velocity) is proportional to the impressed force and takes place in the same direction.

Third Law
Action and reaction are equal and opposite.

Consequences of the Second Law: Definition of the Newton
A mathematical expression of the Second Law is

$$F \alpha \frac{d(mv)}{dt}$$

where m = mass
and   v = velocity
For an object of constant mass, the mass may be removed from the differentiation with respect to time

$$F \alpha m \frac{dv}{dt}$$

$$\alpha \, ma$$

$$\text{since } a = \frac{dv}{dt}$$

or F = kma

k = constant of proportionality.
We let k = 1 and define our units accordingly:

$$\mathbf{F = ma}$$

The Newton is defined as that force which when acting on a mass of 1 kg gives it an acceleration of $1 \, m \, s^{-2}$. **Hence $1 \, N = 1 \, kg \, m \, s^{-2}$**
This convention also enables us to modify the Second Law such that force **equals** rate of change of momentum.

Notes:
1   An important other consequence of F = ma enables us to calculate the weight of an object. When the acceleration concerned is that of gravity, the force concerned is weight.
   weight = mass × acceleration due to gravity
   (Thus 1 kg weighs g Newtons.)
2   Sometimes, at 'A' level, we are not concerned with a constant mass

which is accelerating but with a changing mass moving at a constant velocity. A typical example of this is sand falling from a hopper onto a conveyor belt. Here the mass of the sand moving at a constant velocity (excluding the initial turbulence when it falls onto the belt) is increased.

$$F = \frac{d(mv)}{dt}$$

becomes

$$F = v\frac{dm}{dt}$$

We are not concerned, at 'A' level, with cases where both mass and velocity change simultaneously.

3 Students are reminded that experimental verification of $F = ma$ is required and can be found in any standard 'O' level text. Basically one loads a dynamics trolley with various masses and whilst subjecting it to various forces, measures the corresponding accelerations produced.

Examples

1 A force of 500 N acts upon a body of mass 5 kg. If the resistive force of friction is equal to twice the weight of the body, find the net acceleration produced, assuming $g = 10 \, m \, s^{-2}$.

$$\text{Weight of body} = mg$$
$$= 50 \, N$$
$$\therefore \text{ Frictional force} = 100 \, N$$

$$\therefore \text{ Net force acting on body} = 500 \, N - 100 \, N$$
$$= 400 \, N$$

$$\text{Using } F = ma$$

$$a = \frac{400}{5} = 80 \, m \, s^{-2}$$

Net acceleration = 80 m s⁻²

2 Sand falls vertically at a rate of 0·8 kg s⁻¹ onto a conveyor belt moving with a steady horizontal velocity of 0·25 m s⁻¹. Calculate the force required to keep the belt moving at this velocity.

$$\text{Force required} = v\frac{dm}{dt}$$

$$= 0 \cdot 8 \times 0 \cdot 25$$
$$= 0 \cdot 2 \, N$$

Force required = 0·2 N

Linear momentum and its conservation

Momentum (or more properly linear momentum) is defined as the product of mass and linear velocity.

momentum = mass × velocity
Unit: kg m s^{-1}   **Dim: MLT^{-1}**

The Principle of Conservation of Linear Momentum states: 'When two or more bodies collide the total momentum before impact measured in a given direction is equal to the total momentum after impact measured in the same direction.'

Note: Direction is very important since momentum is a vector quantity.

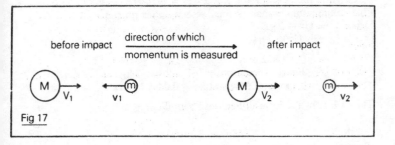

Fig 17

In Fig. 17.
Total momentum before impact $= MV_1 - mv_1$
Total momentum after impact $\;\;= MV_2 + mv_2$
$$MV_1 - mv_1 = MV_2 + mv_2.$$

Notes:
1   If the objects coalesce then they obviously move off with a common velocity.
2   In the case of identical masses, a total transfer of momentum occurs with the incoming object brought to rest and the outgoing object leaving with the velocity which the incoming object had on impact.
3   Experimental verification of the Principle of Conservation of Linear

Momentum is made using the apparatus of $F = ma$. Two trolleys are used with velocity of each measured before and after coalescence.

Example

A body of mass 100 g falls freely from rest at a height of 30 cm vertically above a horizontal surface and rebounds to a height of 20 cm. Find the loss of momentum. ($g = 10 \, \text{m s}^{-2}$.)

Original momentum = mass × original velocity before impact.

$$\text{Using } v^2 = u^2 + 2gs$$

(u = 0 here)
$$= 2 \times 10 \times 0{\cdot}3$$

$$= 6$$

$$v^2 = 2{\cdot}449 \, \text{m s}^{-1}$$

$\therefore$ Original momentum = $0{\cdot}1 \times 2{\cdot}499 = \mathbf{0{\cdot}2449 \, kg \, m \, s^{-1}}$

Final momentum = mass × final velocity after impact

Using $v^2 = U^2 - 2gs$

(v = 0 here) $U^2 = 2gs = 2 \times 10 \times 0{\cdot}2 = 4$

$$\therefore v = 2 \, \text{m s}^{-1}$$

Final Momentum = $0{\cdot}1 \times 2 = 0{\cdot}2 \, \text{kg m s}^{-1}$

$\therefore$ **Loss of Momentum = $0{\cdot}0449 \, \text{kg m s}^{-1}$**

This is lost in the form of heat and sound energy.

Practice questions

1   A barrel of mass 50 kg is being pulled up a smooth plane inclined at 30° to the horizontal by a rope which is parallel to the surface of the plane. If the tension in the rope is 300 N, find the acceleration of the barrel. (Assume $g = 10 \, \text{m s}^{-2}$.)

2   600 biscuits per minute are fed vertically onto a conveyor velt moving at a steady horizontal velocity of $0{\cdot}1 \, \text{m s}^{-1}$. Assuming that the force required to drive the belt is $0{\cdot}01$ N whilst the process occurs, calculate the mass of each biscuit.

3   An object of mass 3 kg moving with a velocity of $3 \, \text{m s}^{-1}$ collides head on with an object of mass 2 kg moving with a velocity of $0{\cdot}75 \, \text{m s}^{-1}$ in the opposite direction. Calculate the velocity with which they both move off after they coalesce and state the direction of motion.

5   WORK, ENERGY AND POWER

Work

Whenever the point of application of a force moves, work is done. Work is defined as the product of force and distance moved in the direction of the force.

A general mathematical expression for work done can be derived as follows.

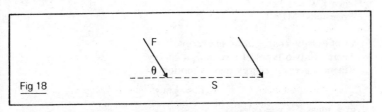

Fig 18

Imagine a force F moves its point of application through a distance S, the angle between the two directions being θ, then

$$\text{work done} = F\,S\,\text{Cos}\,\theta.$$

Students are advised to learn this equation which reduces to a simpler form in two special cases:

1   When $\theta = 0°$, $\text{Cos}\,\theta = 1$ and the expression reduces to $F\,S$
2   When $\theta = 90°$, $\text{Cos}\,\theta = 0$ no work is done since the force has no component in the direction of motion.

The unit of work is the Joule (J) which is defined as the work done when a force of 1 N moves its point of application through a distance of 1 m in its own direction.

Hence $1\,\text{J} = 1\,\text{N\,m}$.

Dimensions of Work: $M\,L^2\,T^{-2}$.

Energy

Energy is defined as the capacity to do work.

It thus has the same units as work and may be thought of as a work reservoir.

Power

Power is defined as the rate of doing work

$$P = \frac{dW}{dt} = \frac{d(F\,S\,\text{Cos}\,\theta)}{dt} = F\,v\,\text{Cos}\,\theta$$

P = power   W = work   t = time

(θ is usually 0°)

The unit of power is the WATT (W) which is defined as a rate of working of 1 J every second.

   Hence $1\,W = 1\,J\,s^{-1}$

   Dimensions: $ML^2\,T^{-3}$

Types of energy, conservation of energy

   There are two types of mechanical energy:

1 **Kinetic Energy** – energy due to motion

$$\text{K.E.} = \tfrac{1}{2}\,m\,v^2$$

   m = mass   v = velocity

2 **Potential Energy** – energy due to position or state.

   (a) The potential energy of an object due to the earth's

$$\text{P.E.} = m\,g\,h$$

   m = mass   g = acc. of grav.   h = height

   (b) The potential energy of a stretched wire or spring – strain energy – given by $\tfrac{1}{2}k\,x^2$

   k = Hooke's Const.   x = extension

The Principle of Conservation of Energy states 'Energy can neither be created nor destroyed but may be converted from one form into another.'

Notes

1 We include mass as a form of energy given by the Einstein Mass Equivalence Relationship $E = m\,c^2$

   E = energy of mass m   c = velocity of light 'in vacuo'

2 When dealing with energy transformations we adopt the following approaches:

   either we ignore all energy losses e.g. when a stone falls we assume that all P E changes into K E without any losses due to air resistance.

   or we assume that all energy is usefully utilised in doing work e.g. the K E of a bullet entering a target is used up in penetration.

Examples

1 Calculate the power required to drive a motor car at a steady velocity of $30\,km\,hr^{-1}$ if the resistive forces equal 200 N.

For a steady velocity:
Force provided by engine = Resistive Forces

$$= 200 \, \text{N}$$

$$30 \, \text{km Hr}^{-1} = \frac{30 \times 1000}{3600} = \frac{25}{3} \, \text{m s}^{-1}$$

$$\therefore \; P = F = 200 \times \frac{25}{3} = 1 \cdot 667 \, \text{kW}$$

2 A bullet of mass 5 g travelling at $100 \, \text{m s}^{-1}$ hits a target and is brought to rest after travelling 5 cm. Calculate the kinetic energy of the bullet before impact and the average retarding force acting on the bullet when it hits the target.

$$\text{K.E.} = \frac{1}{2} \, \text{m} \, \text{v}^2 = \frac{1}{2} \times \frac{5}{1000} \times 100 \times 100 = \textbf{25 J}$$

Assume all K.E. used in overcoming retarding force

$$\text{Loss of K.E.} = \text{Work done}$$

$$\tfrac{1}{2} \, \text{m} \, \text{v}^2 - 0 = \text{F S}$$

$$25 = F \times \frac{5}{100}$$

$$\therefore \; F = 500 \, \text{N}$$

Practice questions

1 A boy of mass 200 kg climbs a vertical ladder of height 7·5 m, at a uniform rate, in 15 s. Find the power he develops. (Assume g $= 10 \, \text{m s}^{-2}$.)
2 The bob of a pendulum of length 1 m has a mass of 100 g. The bob is struck by a moving object of mass 20 g which coalesces with it and deflects the bob through 30°. Calculate the impact velocity of the moving object. (Assume g $= 10 \, \text{m s}^{-2}$.)

6 MOMENTS AND COUPLE

Moments

The moment of a force is defined as 'The product of its magnitude and the perpendicular distance from the point to the line of action of the force'

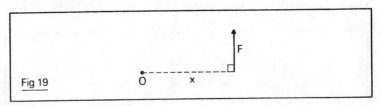

Fig 19

Moment of F about 0 = Fx Nm
Unit: Nm; Dim: $ML^2 T^{-2}$
A moment is the tendency of a force to rotate the body upon which it acts.

Moments can be either clockwise or anti-clockwise depending upon the direction in which they act and they can be added algebraically. Experiment and theory show that the moment of a number of forces about any point is equal to the algebraic sum of the moments of the individual forces about the same point.

In Fig. 20
Resultant moment of forces about $0 = F_1 x_1 - F_2 x_2 + F_3 x_3$.

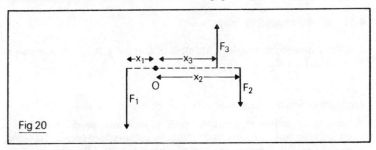

Fig 20

The algebraic sum of the moments of all the forces about any point is zero when the forces are in equilibrium.

Couples

A pair of equal and opposite parallel forces, not acting in the same line, constitute a couple.

The two forces have a turning effect known as the **Moment of The Couple or Torque** defined as:

One force × Perpendicular distance between forces

In Fig. 21
Moment of couple = Fx Nm (not 2Fx!).

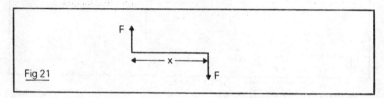

Fig 21

Unit: Nm   Dim: $ML^2 T^{-2}$

Important properties of couples:

1   A couple tends to produce pure rotation

2   A couple can never be replaced by a single force

3   A couple can only be balanced by another couple of equal moment and opposite sign

4   Couples add algebraically

5   Work done by a couple = Moment of Couple × Angle of rotation of couple. (in radians)

Uniform circular motion

When a body moves in a circular path at constant speed its velocity is constantly changing because its direction is continually changing. However, since the speed is constant the **numerical** magnitude of its velocity is also constant – a fact which should be borne in mind because often the terms speed and velocity in such cases are mixed and interchanged.

In circular motion, we measure angles in radians: r

Important relationships
1   $x = r\theta$

where x = distance along circumference
    r = radius of circle
    θ = angle subtended by circumference

2 $\omega = \dfrac{d\theta}{dt}$

i.e. Angular velocity = Rate of change of angular displacement
(measured in r s^{-1})

3 $v = r\omega$
  ω = angular velocity
  r = radius of circle
  v = linear velocity around circumference

4 Tangential acceleration = 0

5 Radial acceleration = $\dfrac{v^2}{r}$ or $r\omega^2$ **towards centre**

Note: The acceleration is directed towards the centre of the circle and is known as the **centripetal acceleration** (the associated force being the centripetal force); the Newtonian reaction is the centrifugal force. Thus:

centripetal force = real force, towards centre
centrifugal force = Newtonian reaction, away from centre

Application of centrifugal force: the centrifuge.

The conical pendulum

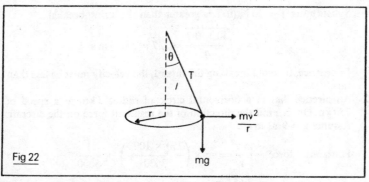

Fig 22

Consists of a mass m on the end of a string, length *l*, orbiting at angle $\theta°$ in a horizontal circle of radius r.

Let the linear velocity of the mass be v.
Resolving vertically

(a) $$T \cos \theta = mg$$

Resolving horizontally

(b) $$T \sin \theta = \frac{mv^2}{r}$$

Dividing (b) by (a)

$$\tan \theta = \frac{v^2}{rg} = \frac{v^2}{gl\sin\theta}$$

As v increased so does θ.

Examples

1 A toy cart of mass 4 kg and at the end of a string 0·7 m long, moves in a circle on a table. If the breaking strength of the string is 40 N, calculate the maximum speed of the cart.

Let the tension be T, cart mass be m and its velocity be v whilst its radius of orbit is r.
Then

$$T = \frac{mv^2}{r}$$

$$v = \sqrt{\frac{Tr}{m}}$$

Substituting T = 40 N (if T is greater then the string breaks).

$$\therefore \ v = \sqrt{\frac{40 \times 0\cdot7}{4}} = \sqrt{7} = 2\cdot646 \text{ m s}^{-1}$$

In practice, to avoid breaking the thread, the velocity must be less than this.

2 An aircraft flies in a horizontal circle of radius 5 km at a speed of 750 km Hr^{-1}. Find the direction of the resultant force on the aircraft. Assume $g = 9\cdot81$ m s^{-2}

Horizontal force $= \dfrac{mv^2}{r} = m \times \left(\dfrac{750 \times 1000}{3600}\right)^2 \times \dfrac{1}{5000}$

$$= 8\cdot681 \ m$$

Vertical force $\quad = mg = 9\cdot81 \ m$

These are shown in Fig. 23.

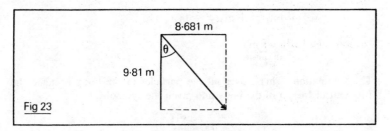

Fig 23

If θ = angle to the vertical

$$\text{Tan } \theta = \frac{8 \cdot 681}{9 \cdot 81}$$

$\therefore \ \theta = 41° \ 31'$

$\therefore \ $ **41° 31' to the vertical**

Rigid body rotation

Moment of inertia

In order to introduce this concept, we consider the kinetic energy of a rotating body.

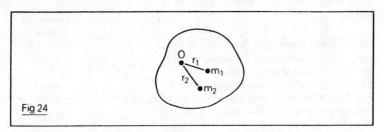

Fig 24

Suppose a rigid body rotates about a fixed axis 0. The body consists of a large number of minute bodies m_1, m_2 ... etc. and since the body is rigid all have the same angular velocity, ω

Total K.E. of body $= \frac{1}{2} m_1 v_1^2 + \frac{1}{2} m_2 v_2^2 + \frac{1}{2} m_3 v_3^2 + \ldots$

But $v_1 = r_1 \omega$, $\qquad v_2 = r_2 \omega$, $\qquad v_3 = r_3 \omega$

$$\therefore = \text{K.E.} = \tfrac{1}{2} m_1 r_1^2 \omega^2 + \tfrac{1}{2} m_2 r_2^2 \omega^2 + \tfrac{1}{2} m_3 r_3^2 \omega^2 \ldots$$

$$= \tfrac{1}{2} \omega^2 (m_1 r_1^2 + m_2 r_2^2 + m_3 r_3^2)$$

$$= \tfrac{1}{2} \omega^2 \sum_{\substack{\text{All particles} \\ \text{in body}}} m \, r^2$$

The summation, $\Sigma \, m \, r^2$, over all the particles in the body is called the Moment of Inertia of the body and given the symbol I.

$$I = \sum_{\substack{\text{all particles} \\ \text{in body}}} m \, r^2$$

Units are $kg\, m^2$; Dimensions: ML^2
Hence:

$$\text{K.E.} = \tfrac{1}{2} I \omega^2$$

We may consider I to be a constant relating K.E. of rotation to the square of angular velocity.

Moments of inertia of some familiar objects
The calculation of the moment of inertia of a body is done using calculus and yields the following results:

(m = mass, *l* = total length, a = radius)

Object	Moment of inertia
Uniform rod about axis through middle	$ml^2/12$
Uniform rod about axis through one end	$ml^2/3$
Ring about axis through centre	ma^2
Circular disc about axis through centre	$ma^2/2$
Solid cylinder about axis perpendicular to middle	$ma^2/2$
Sphere about an axis through middle	$2ma^2/5$

Example
A disc of mass 6 kg has a radius of 10 cm and rotates at an angular velocity of $120\, r\, s^{-1}$. Calculate its rotational kinetic energy.

$$I = \frac{ma^2}{2} = \frac{6 \times 0{\cdot}1^2}{2} = 3 \times 10^{-2} \, kg\, m^2$$

$$\text{K.E.} = \frac{1}{2} I w^2 = \frac{3 \times 10^{-2} \times 120^2}{2}$$

$$K.E. = 216 \text{ J}$$

Angular Momentum
Angular Momentum is defined as 'Moment of Momentum'.
Units are $kg\, m^2\, s^{-1}$. Dimensions: $ML^2\, T^{-1}$.

To find the angular momentum of a rotating body consider again Fig. 24.
The angular momentum of $m_1 = r_1 \times m_1 v_1$
$$= r_1 \times m_1 r_1 \omega$$
$$= m_1 r_1^2 \omega \qquad kg\, m^2\, s^{-1}$$

Total angular momentum of whole body $= \omega \sum_{\substack{\text{all particles} \\ \text{in body}}} m\, r^2 = \omega I$

$\therefore$ **Angular Momentum $= I\, W$** $\qquad kg\, m^2\, s^{-1}$

'I' may be considered to be a constant relating angular momentum to angular velocity.

Example
A flywheel of moment of inertia $0.35\, kg\, m^2$ rotates steadily at 19 revolutions every second. Calculate its angular momentum.

Angular Momentum $= I\, \omega$

Now 19 revs every second $= 19 \times 2\pi = 19 \times \dfrac{44}{7}\, r\, s^{-1}$

$\therefore$ Angular Momentum $= 0.35 \times 19 \times \dfrac{44}{7} = 41.8\, kg\, m^2\, s^{-1}$

Angular Momentum $= 41.8\, kg\, m^2\, s^{-1}$

Conservation of Angular Momentum

As with linear momentum, angular momentum is conserved

Example

A horizontal disc, moment of inertia $7.3 \times 10^{-4}\, kg\, m^2$ rotates about a vertical axis making 100 revolutions per minute. A small piece of wax falls vertically onto the disc and sticks to it at a distance of 10 cm from the axis. If the number of revolutions per minute is reduced to 80 calculate the mass of the wax.
Use the Conservation of Angular Momentum.

Original angular momentum $=$ that due to disc

$$= I\,\omega$$

$$= 7{\cdot}3 \times 10^{-4} \times \frac{100}{60} \times \frac{44}{7}$$

$$= 7{\cdot}648 \times 10^{-3}\ \text{kg m}^2\,\text{s}^{-1}$$

Final angular momentum   = that due to disc + that due to wax
$$= I\,\omega + \omega\,m\,r^2$$

($I\,\omega = \omega$, $m\,r^2 = \omega\,m\,r^2$ since only one body.)

$$= 7{\cdot}3 \times 10^{-4} \times \frac{80}{60} \times \frac{44}{7} + \frac{80}{60} \times \frac{44}{7} \times m \times 0{\cdot}1 \times 0{\cdot}1$$

$$= 6{\cdot}118 \times 10^{-3} + 8{\cdot}381 \times 10^{-2}\,m \quad \text{kg m}^2\,\text{s}^{-1}$$

But since Angular Momentum is conserved:
$$7{\cdot}648 \times 10^{-3} = 6{\cdot}118 \times 10^{-3} + 8{\cdot}381 \times 10^{-2}\,m$$
$$\therefore\ m = 1{\cdot}826 \times 10^{-2}\ \text{kg}$$
Mass = $1{\cdot}826 \times 10^{-2}$ kg   or   18·26 g.

Couple acting on a rotating body
Consider again Fig. 24.

Force acting on $m_1 = m_1\,a = m_1\dfrac{dv}{dt} = m_1\dfrac{d(r_1\,\omega)}{dt}$

$$= m_1\,r_1\frac{d\omega}{dt}$$

$$= m_1\,r_1\frac{d(d\theta/dt)}{dt} = m_1\,r_1\frac{d^2\theta}{dt^2}\,\text{N}$$

$\therefore$ Moment of Couple = Force × Perpendicular distance

$$= m_1\,r_1{}^2\frac{d^2\theta}{dt^2}\,\text{N m}$$

$\therefore$ Moments of Couple acting on whole body $= \displaystyle\sum_{\substack{\text{All particles}\\ \text{in body}}} m_1\,r_1{}^2\frac{d^2\theta}{dt^2}$

$$= I\frac{d^2\theta}{dt^2}$$

$$\therefore\ C = I\frac{d^2\theta}{dt^2}$$

Moment of Couple acting = M of I × Angular Acceleration or **torque.**

Example

A uniform circular disc of mass 20 kg and radius 0·4 m is mounted on a horizontal axle of radius 0·01 m and negligible mass. Ignoring friction calculate the angular acceleration if a force of 40 N is applied tangentially to the axle.

Torque applied = Force applied to axle × Axle radius

$$= 40 \times 0 \cdot 01 = 0 \cdot 4 \, N \, m$$

M of I of disc $= \dfrac{Ma^2}{2} = \dfrac{20 \times 0 \cdot 4^2}{2} = 1 \cdot 6 \, kg \, m^2$

Angular acceleration $= \dfrac{Torque}{M \, of \, I} = \dfrac{0 \cdot 4}{1 \cdot 6} = 0 \cdot 25 \, r \, s^{-2}$

Angular Acceleration $= 0 \cdot 25$ **radians sec^{-2}**

Radius of gyration

We can express the moment of inertia as:

$$\sum m \, r^2 = M \, k^2$$

where M = total mass of body

k = radius of gyration, in metres.

In effect we are adopting a 'centre of mass' approach.

Example

To find the radius of gyration of a uniform rod about an axis through its middle.

$$I = \frac{m \, l^2}{12}$$

But

$$I = M \, k^2$$

$$\therefore \; k^2 = \frac{l^2}{12} \quad \text{or} \quad k = \frac{l}{\sqrt{12}} m$$

Theorem of parallel axes

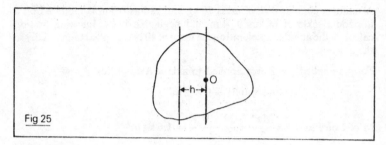

Fig 25

Theorem states

$$I = I_G + M h^2$$

Where I_G = M of I about an axis passing through centre of mass, 0.
   I = M of I about a parallel axis, a distance h from the axis through 0.
   M = Mass of body.

Example

Given that the moment of inertia of a uniform rod about an axis through its middle is $ml^2/12$, show that its moment of inertia about an axis through one end is $ml^2/3$.

Since the rod is uniform, its centre of mass lies on a line through its midpoint.
Hence $I_G = ml^2/12$.
The axis through the end is parallel to this at a distance $l/2$ away.
The M of I about the end is thus:

$$I = I_G + M h^2$$

$$= \frac{ml^2}{12} + m\left(\frac{l}{2}\right)^2 = \frac{ml^2}{12} + \frac{ml^2}{4} = \frac{ml^2}{3}$$

Comparison between linear & rotational motion

	linear	rotational
Kinetic Energy (J)	$\frac{1}{2} m v^2$	$\frac{1}{2} I W^2$
Momentum (kg m s^{-1})	m v	I W
Force (N)	m a	I $d^2\theta/dt^2$

It can be seen that
I is the rotational counterpart of mass.

Angular velocity is the rotational counterpart of linear velocity.
Angular acceleration is the rotational counterpart of linear acceleration.
The idea can be extended to give 'Rotational' Equations of Motion:

$$\omega = \omega_0 + \alpha t$$
$$\omega^2 = \omega_0{}^2 + 2\alpha\theta$$
$$\theta = \omega_0 t + \tfrac{1}{2}\alpha t^2$$

ω_0 = initial angular velocity
ω = final angular velocity
α = angular acceleration
θ = angular displacement
t = time

Kinetic energy and acceleration of rolling objects

When an object rolls on a plane it is rotating as well as translating along the plane. It therefore has translational and rotational energy. In order to simplify calculations, we consider the object to rotate about an axis at the point of contact. We can then arrive at a general formula for total kinetic energy.

In order to calculate acceleration we use the approach that loss of potential energy is equal to the gain of kinetic energy and relate linear velocity to acceleration using a linear equation of motion. It can be shown that the acceleration of an object is smaller when rolling than when sliding.

Example A
To calculate the kinetic energy and acceleration of a solid cylinder rolling down a plane, inclined at $\alpha°$ to the horizontal.

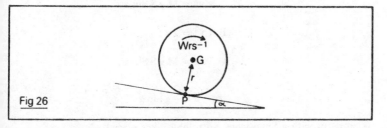

Fig 26

Let the mass of the cylinder be M kg and its radius be r m. At any instant the line of contact through P is at rest and we consider the cylinder to be rotating about this line.

If I_G = M of I about axis through the centre of mass (along G) and I_p = M of I about line of contact

$$I_p = I_G + M h^2$$
$$I_p = I_G + Mr^2$$

But

$$K.E. = \frac{1}{2} I_p \ \omega^2 = \frac{1}{2} I_G \ \omega^2 + \frac{1}{2} Mr^2 \ \omega^2$$

But $v = r\omega$

$$\therefore K.E. = \frac{1}{2} I_G \omega^2 + \frac{1}{2} M v^2$$

TOTAL K.E. = K.E. of Rotation + K.E. of Translation
(about axis through centre of mass)
This is **always** true
We can simplify this equation by substituting for I_G and N.

$$K.E. = \frac{1}{2} \frac{Mr^2}{2} \omega^2 + \frac{1}{2} Mr^2 \omega^2 = \frac{3}{4} Mr^2 \omega^2$$

$$or \quad \frac{3}{4} Mv^2$$

In order to calculate the acceleration of this cylinder, suppose it rolls from rest through a distance S metres

$$Loss \ of \ P.E. = M g S \ Sin \ \alpha$$

$$\therefore \frac{3}{4} Mv^2 = M g S \ Sin \ \alpha$$

$$But \ v^2 = 2 a s$$

$$\therefore \frac{3}{4} M 2 a s = M g S \ Sin \ \alpha$$

$$or \quad a = \frac{2}{3} g \ Sin \ \alpha$$

(Compares with g Sin α when no rolling takes place.)

Practice questions

1 Couples of 8 nm clockwise and anticlockwise act on a body. Find the resultant couple and comment upon the result.

2 A 2 kg stone at the end of a 1 m string is made to describe a vertical circle at a constant speed of 4 m s^{-1}. Calculate the tension in the string when the stone is at the bottom of the circle, assuming g = 10 m s^{-2}.

3 The bob of a metre long pendulum has a mass of 20 g and is made to describe a horizontal circular orbit of radius 50 cm. Calculate the speed of the bob and the tension in the string, assuming g = 10 m s^{-2}.

4 A sphere of mass 5 kg rotates about an axis through its centre such that the linear velocity of a point on its surface is 2 m s^{-1}. Calculate the kinetic energy of rotation of the sphere. (Hint: The problem isn't difficult – write down the expression for rotational K.E. and substitute for I in the expression. Remember that v = r w.)

5 Calculate the angular momentum of a horizontal disc, moment of inertia 7·3 × 10^{-4} kg m^2 when it rotates about a vertical axis through its centre making 100 revolutions per minute.

6 A uniform circular disc, radius a m, rotates at a steady rate about an axis through its centre such that it is rotating in a horizontal plane. A circular blade, whose axis of rotation is parallel to and co-incident with the axis of the disc cuts the disc and reduces its radius to half of its initial value. Assuming that angular momentum is conserved, calculate the ratio of the final to the initial angular velocity.

7 The moment of inertia of a solid flywheel about its axis is 0·2 kg m^2. It is set in rotation by applying a tangential force of 20 N with a cable wound round the circumference. If the radius of the wheel is 0·2 m, calculate the angular acceleration of the flywheel.

8 Calculate the radius of gyration of a uniform rod about an axis through one end.

9 Given that the M of I of a sphere about an axis through its centre is 2/5 M a^2 (M = mass, a = radius), use the theorem of parallel axes to find the M of I about an axis passing through its circumference.

10 Repeat example A for a sphere of radius a metres, mass M kg.

7 SIMPLE HARMONIC MOTION (S.H.M.)

Simple Harmonic Motion is a very important type of motion in physics. Apart from being the easiest kind of motion to describe after straightforward linear motion, it is important because it is a common type of motion found in **oscillating systems**. It can also be a first approximation to more complicated types of motion. However, one cannot fall into the trap of using it universally – indeed if all the atoms on a crystal lattice vibrated with simple harmonic motion the phenomenon of linear expansivity would not arise.

A particle moves with simple harmonic motion when its acceleration towards a fixed point in its path varies directly with its distance from that fixed point, measured along the path.

i.e. **acceleration = − constant × displacement**

Equation of simple harmonic motion

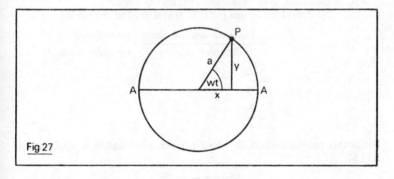

Fig 27

Consider Fig. 27 which shows a point P traversing the circumference of a circle of radius a with a constant linear speed v.

The relationship between v and the angular velocity ω is:

$$\omega = \frac{v}{a}$$

The projection of P onto the horizontal diameter A A^1 moves with S.H.M. From Fig. 27

$$x = a \cos \omega t$$

$$\therefore \ v = \frac{dx}{dt} = -a\,\omega\,\text{Sin}\,\omega t$$

$$\therefore \ a = \frac{dv}{dt} = -a\,\omega^2\,\text{Cos}\,\omega t$$

Thus

$$\mathbf{a = -\omega^2 x}$$

this is the mathematical equation of s.h.m.

acceleration $= -(\text{ang. vel})^2 \times$ **displacement**

Note: It can equally be derived by considering a projection on the vertical diameter.

Amplitude is defined as the maximum value of x.

Again considering Fig. 27.

The period of the motion, T, is defined as 'The time taken for one complete cycle of the motion to occur'. Thus it is the time taken for the projection on the horizontal diameter to go from A to A^1 and back again to A.

But, in that time, P has gone once around the circle.

$$\therefore \ \text{Period T} = \text{Time for P to travel around the circle}$$

$$= \frac{\text{Distance around circle} \quad \text{(in radians)}}{\text{Angular velocity} \quad \text{(in radians sec}^{-1})}$$

$$= \frac{2\pi}{\omega}$$

$$\mathbf{T = \frac{2\pi}{\omega}}$$

From this can be derived the third mathematical equation applying to S.H.M.

$$\text{Since } a = -\omega^2 x$$

Omitting the minus sign, which indicates a restoring force:

$$\omega = \sqrt{\frac{a}{x}}$$

$$\therefore \ T = 2\pi\sqrt{\frac{x}{a}}$$

This equation enables us to immediately find the period if we know the acceleration corresponding to a particular displacement.

To find the period of a simple pendulum, oscillating with S.H.M.

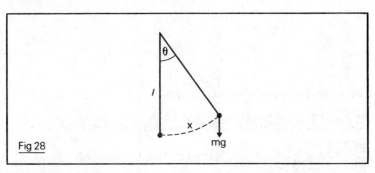

Fig 28

In Fig. 28

Let l = length of pendulum

   m = mass of bob

   θ = angular displacement (in radians)

   T = tension in string

Forces acting on the bob: Tension, T

                                   Weight, mg

The component of the weight along the string is balanced by the tension

    $T = m\,g\,\cos\theta.$

The component of the weight perpendicular to the string acts to decrease x and is a restoring force.

$$\therefore\ F = -m\,g\,\sin\theta$$

But

$$F = m\frac{d^2x}{dt^2}$$

$$\therefore\ m\frac{d^2x}{dt^2} = -m\,g\,\sin\theta$$

or

$$\frac{d^2x}{dt^2} = -g\,\sin\theta$$

Now if θ is very small ($< 5°$) then $\sin\theta \approx \theta \approx \dfrac{x}{l}$

$$\therefore\ \frac{d^2x}{dt^2} = -\frac{g}{l}x$$

or **acc = −constant × displacement**

This therefore shows that the motion is simple harmonic with $\omega^2 = g/l$

Now

$$T = \frac{2\pi}{\omega} = \frac{2\pi}{\sqrt{g/l}}$$

$$T = 2\pi\sqrt{\frac{l}{g}}$$

The motion is simple harmonic, of period $T = 2\pi\sqrt{l/g}$.

Example

Show that liquid oscillating in a U tube oscillates with S.H.M. and calculate the period of oscillation of a length of liquid of 1 m, assuming $g = 10 \text{ m s}^{-2}$.

Let the total length of liquid be l m and let it be contained in a U tube of cross-sectional area A m², its density being ρ kg m⁻³.

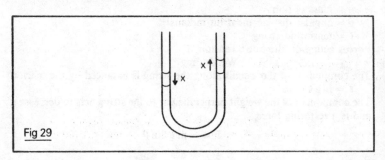

Fig 29

We ignore frictional effects.

Suppose that the free surface in one arm is depressed by x m, causing a head 2x m.

Restoring Force = − Mass of liquid × g
= − Density × Volume × g
= − 2ρ Axg

But

$$F = m\frac{d^2x}{dt^2}$$

The mass upon which this force acts is the **total** mass of liquid m = l Aρ

$$\therefore \; l \, A \, \rho \, \frac{d^2x}{dt^2} = -2 \, \rho \, A \, x \, g$$

$$\frac{d^2x}{dt^2} = -\frac{2 \, g}{l} \, x$$

This is S.H.M., with
$$\omega^2 = \frac{2 \, g}{l}$$

Now

$$T = \frac{2\pi}{\omega} = \frac{2\pi}{\sqrt{2 \, g/l}} = 2\pi \sqrt{\frac{l}{2 \, g}}$$

Substituting $l = 1$ m, $g = 10 \, \text{m s}^{-2}$

$$T = 2\pi \sqrt{\frac{l}{20}} = 1 \cdot 405 \text{ s}$$

Period = 1·405 s

Velocity during S.H.M.

We have seen:

$$V = -a \, \omega \sin \omega t$$

Now refer again to Fig. 27. It is necessary to write the expression in terms associated with the motion of the projection of P on the horizontal axis.

But

$$Sin \, \omega \, t = \frac{y}{a}$$

Now

$$a^2 = y^2 + x^2$$
$$\therefore \; y = \pm \sqrt{a^2 - x^2}$$

Hence

$$Sin \, \omega \, t = \pm \sqrt{\frac{a^2 - x^2}{a}}$$

So

$$V = -a\omega \, \text{Sin} \, \omega t = \pm \frac{a\omega\sqrt{a^2 - x^2}}{a}$$

$$V = \pm\omega\sqrt{a^2 - x^2}$$

The + and − signs tell us that the value of the velocity can be either away from or towards the rest position.

The numerical values of the maximum and minimum velocities are $V = V_{max}$ when $x = 0$ i.e. **passing through rest position**

$$V_{max} = \omega a \ \text{m s}^{-1}$$

$$V = V_{min} \text{ when } x = a$$

i.e. at the end of the oscillation when the displacement is equal to the amplitude

$$V_{min} = 0 \, \text{m s}^{-1}$$

Note: This relationship between displacement and velocity is to be expected because the phase difference between them is 90°.

Acceleration during S.H.M.

We have seen:

$$a = -\omega^2 x.$$

The numerical values of maximum and minimum accelerations are

$$a = a_{max} \text{ when } x = a$$

i.e. when the displacement is equal to the amplitude

$$a_{max} = \omega^2 a \ \text{m s}^{-2}$$

$$a = a_{min} \text{ when } x = 0$$

i.e. passing through rest position

$$a_{min} = 0 \, \text{m s}^{-2}$$

Note: This relationship between acceleration and displacement is to be expected for the following two reasons:
1   Acceleration and displacement are in phase.
2   Acceleration and velocity are 90° out of phase.

Additionally, without resort to equations, when velocity is a maximum the acceleration must be zero.

The relationship between displacement, velocity and acceleration may be seen in Fig. 30.

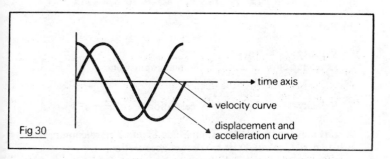

time axis

velocity curve

displacement and acceleration curve

Fig 30

Note: We have shown the amplitude to be the same for simplicity. In fact, of course:

Amplitude of Velocity Plot = $a\omega \times$ Displacement Amp

and

Amplitude of Acceleration Plot = $a\omega^2 \times$ Displacement Amp

Examples

1   An object moving with S.H.M. has an amplitude of 0·03 m and a frequency of 30 Hz. Calculate its acceleration and velocity at the middle and end of an oscillation.

(a) **Middle**

Acceleration = 0 m s^{-2}
Velocity   = $\omega a \text{ m s}^{-1}$, the maximum value
a   = $0·03$ m

In order to find ω we use the fact that:

$$T = \frac{1}{f} \quad \text{and} \quad T = \frac{2\pi}{\omega}$$

$$\therefore \ \omega = 2\pi f$$

where f = frequency in Hz

(This is the familiar relationship from the physics of vibrations and waves.)

$$\therefore \quad \omega = 60\pi \, \mathrm{r \, s^{-1}}$$
$$\text{Hence velocity} = 0{\cdot}03 \times 60\pi = 1{\cdot}8\pi \, \mathrm{m \, s^{-1}}$$
$$\textbf{Velocity} = \textbf{1·8}\pi \, \textbf{m s}^{-1}, \quad \textbf{Acceleration} = \textbf{0 m s}^{-2}$$

(b) **End**

$$\text{Velocity} \quad = 0 \, \mathrm{m \, s^{-1}}$$
$$\text{Acceleration} = \omega^2 a \, \mathrm{m \, s^{-2}}, \text{ the maximum value}$$
$$= (60\pi)^2 \times 0{\cdot}03$$
$$= 108\pi^2 \, \mathrm{m \, s^{-2}}$$
$$\text{Velocity} \quad = 0 \, \mathrm{m \, s^{-1}}, \quad \textbf{Acceleration} = 108\pi^2 \, \mathrm{m \, s^{-2}}$$

2   A body moves with S.H.M. of amplitude 40 mm. If its maximum velocity is $0{\cdot}16 \, \mathrm{m \, s^{-1}}$, calculate its maximum acceleration.

$$V_{max} = \omega a$$

Substituting for V_{max} and a:

$$\omega = \frac{0{\cdot}16}{0{\cdot}04} = 4 \, \mathrm{r \, s^{-1}}$$

Now

$$a_{max} = \omega^2 a$$
$$= 4^2 \times 0{\cdot}04$$
$$= 0{\cdot}64 \, \mathrm{m \, s^{-2}}$$
$$\textbf{Maximum acceleration} = \textbf{0·64 m s}^{-2}$$

Energy of a particle executing S.H.M.

Kinetic Energy
We use

$$\text{K.E.} = \tfrac{1}{2} m v^2$$

but

$$v = \omega \sqrt{a^2 - x^2}$$
$$\therefore \quad v^2 = \omega^2 (a^2 - x^2)$$

Hence

$$\textbf{K.E.} = \tfrac{1}{2} \, \textbf{m} \, \omega^2 \, (\textbf{a}^2 - \textbf{x}^2)$$

The minimum (zero) K.E. is hence when $x = a$ which is what we would expect because $v = 0$ when $x = a$.

The maximum K.E. is hence when $x = 0$. Again this is to be expected because v is a maximum when $x = 0$.

$$\therefore \text{ K.E. max} = \tfrac{1}{2} m \omega^2 a^2$$

Potential Energy

We have seen that the maximum K.E. occurs at zero displacement i.e. in equilibrium. At this point it must have no potential energy and the total energy is thus the maximum K.E.

$$\therefore \text{ Total Energy} = \tfrac{1}{2} m \omega^2 a^2$$

In order to find the instantaneous energy we use the **Conservation of Energy**.

$$\text{Total Energy} = \text{P.E.} + \text{K.E.}$$

$$\therefore \text{ P.E.} = \text{Total Energy} - \text{K.E.}$$

$$= \tfrac{1}{2} m \omega^2 a^2 - \tfrac{1}{2} m \omega^2 (a^2 - x^2)$$

$$\text{P.E.} = \tfrac{1}{2} m \omega x^2$$

The minimum (zero) P.E. occurs when $x = 0$, at equilibrium. The maximum P.E. occurs when $x = a$. This is to be expected since its velocity is then zero and it has no K.E.

$$\text{P.E. max} = \tfrac{1}{2} m \omega^2 a^2$$

In S.H.M. energy continually changes from potential to kinetic and back to potential.

The energy changes are represented in Fig. 31.

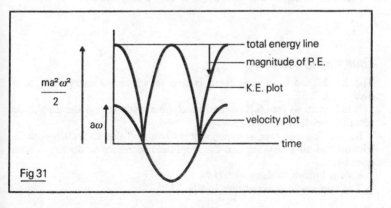

Fig 31

Example

A particle of mass 2 g moves with S.H.M. of period π s and amplitude 10 cm. **Calculate** the K.E. of the particle when passing through its equilibrium position and when it is 5 cm from the equilibrium position.

Use

$$\text{K.E.} = \tfrac{1}{2}\,m\,\omega^2\,(a^2 - x^2)$$

1 At equilibrium position

$$x = 0\,\text{m}$$
$$a = 10\,\text{cm} = 10^{-1}\,\text{m}$$
$$m = 2\,\text{g} = 2 \times 10^{-3}\,\text{kg}$$
$$T = \frac{2\pi}{\omega} = \pi$$
$$\therefore\ \omega = 2\,\text{r s}^{-1}$$
$$\therefore\ \text{K.E.} = \tfrac{1}{2} \times 2 \times 10^{-3} \times 4 \times 10^{-2}$$
$$= 4 \times 10^{-5}\,\text{J}$$

2 5 cm from this position

$$x = 5\,\text{cm} = 5 \times 10^{-2}\,\text{m}$$
$$\text{K.E.} = \tfrac{1}{2} \times 2 \times 10^{-3} \times 4 \times (10^{-2} - 25 \times 10^{-4})$$
$$= 3 \times 10^{-5}\,\text{J}$$

Kinetic Energy $= 4 \times 10^{-5}$ J, $\ 3 \times 10^{-5}$ J respectively

Damping of S.H.M.

The S.H.M. we have considered is that in which no energy has been **dissipated**.

It is known as **free** S.H.M. and its characteristic is that the amplitude **remains** constant as time goes by.

In practice, however, some energy is dissipated due to factors such as **friction** and air resistance and the amplitude decreases or decays as time **goes** by.

This is known as **damped** S.H.M.

The two motions are shown in Fig. 32.

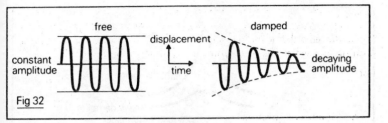

Fig 32

The damping may be of 3 types:

Slight – as shown in Fig. 32 where the amplitude decays gradually.

Critical – the resistance is just sufficient to prevent oscillation, but not so great that the return to the rest position is indefinitely delayed. One quarter of an oscillation is executed. See Fig. 33.

Heavy – the resistance is so great that the return to the rest position is very slow. See Fig. 33.

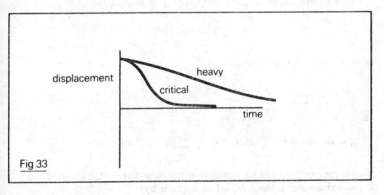

Fig 33

Vibrating systems

All vibrating systems have a frequency at which they vibrate if they are set in oscillation. This is known as the system's **natural frequency**. If some external agent forces the body to vibrate at some frequency other than the natural frequency then the body is said to undergo **forced vibrations** at the forcing frequency.

If the forcing frequency is changed until it matches the natural frequency of the body then the amplitude of vibration increases substantially and the vibrating system is said to be in **resonance**. The variation of amplitude with frequency is shown in Fig. 34.

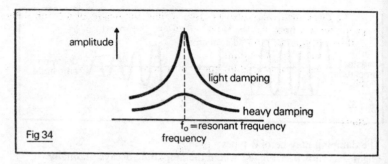

Fig 34

S.H.M. and the determination of g

(1) The Simple Pendulum

This is a standard experiment for the determination of the acceleration due to gravity.

For a pendulum of length l, its period T is given by:

$$T = 2\pi \sqrt{\frac{l}{g}}$$

or

$$T^2 = 4\pi^2 \frac{l}{g}$$

i.e.

$$T^2 = \frac{4\pi^2}{g} l$$

This is the equation of a straight line graph passing through the origin. If T^2 is plotted against l, the gradient m is given by:

$$m = \frac{4\pi^2}{g}$$

thus

$$g = \frac{4\pi^2}{m}$$

In practice one measures the period of a pendulum of known length l by timing 20 oscillations. The process is repeated at least five more times and a graph plotted of T^2 against l. The gradient is measured and g calculated.

The experiment may also be used to find the height of a room if the pendulum were suspended from its ceiling.

If h = height of room

l = length of pendulum

x = height of pendulum above floor level

Now

$$h = l - x$$

Since

$$T^2 = 4\pi^2 \frac{l}{g}$$

then

$$T^2 = 4\pi^2 \frac{(h - x)}{g}$$

$$T^2 = \frac{4\pi^2 h}{g} - \frac{4\pi^2 x}{g}$$

(c.f. $y = mx + c$)

Thus if a graph of T^2 against x is plotted, a straight line of negative gradient, not passing through the origin, is found.

$$\text{Gradient} = \frac{4\pi^2}{g}$$

$$\text{Intercept} = \frac{4\pi^2}{g} h \text{ (on T axis)}$$

Hence h

(2) **The Spring**

This will be discussed under 'Elasticity'.

(3) **The 'Chattering' Object**

If any object rests on a surface which descends at a rate of g m s^{-2}, the object will 'leave the surface' and be in free fall.

If the surface is a horizontal platform moving up and down with S.H.M., the object will leave the surface each cycle when the acceleration of the platform becomes g m s^{-2} downwards (i.e. at the bottom of the oscillation). The mass will 'chatter' as it leaves and returns to the surface.

Example

Calculate the minimum frequency of such a surface if the amplitude of motion is 0·1 m.

Now

$$a = (-)\omega^2 x$$
$$= (-)4\pi^2 f^2 x$$

But

$$a = g$$
$$g = 4\pi^2 f^2 x$$

In order that f should be a minimum, x should be a maximum – that is the amplitude.

$$\therefore \ g = 4\pi^2 f^2 a$$

$$f = \frac{g}{4\pi^2 a}$$

Using

$$g = 10 \, m \, s^{-2}, \quad a = 10^{-1} \, m$$

$$f = \frac{10}{4\pi^2 \times 10^{-1}} = 2 \cdot 533 \, Hz$$

Period of oscillation of a rigid body

If a rigid body oscillates about a fixed axis, the oscillations are simple harmonic.

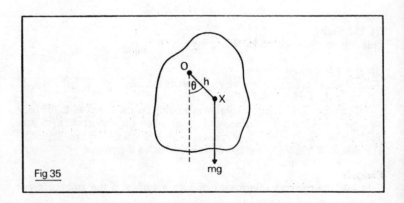

Fig 35

Consider Fig. 35

A body of mass m kg whose centre of mass is at X is suspended at a point O and displaced by an angle θ.

Let the distance between O and X be h m.

The restoring torque is mgh Sin θ.

If θ is small, Sin $\theta \approx \theta$

$\therefore$ Restoring torque = mgh

But

$$\text{torque} = I \frac{d^2\theta}{dt^2}$$

hence

$$-mgh\theta = I \frac{d^2\theta}{dt^2}$$

or

$$\frac{d^2\theta}{dt^2} = -\frac{mgh}{I}\theta$$

This is of the form:

Angular Acc = $-\text{constant}^2 \times$ angular displacement

i.e. **it describes S.H.M.**

Also

$$\omega^2 = \frac{mgh}{I}$$

$$\textbf{The period T} = \frac{2\pi}{\omega} = 2\pi \sqrt{\frac{I}{mgh}}$$

Example

A flywheel of radius 10 cm is pivoted to rotate about a horizontal axis. A piece of wax, mass 1 g, is attached to the edge of the wheel which, after a slight displacement, oscillates with S.H.M. of period π s. Calculate the moment of inertia, assuming that $g = 10 \text{ m s}^{-2}$ and the mass of the wax is insignificant compared with that of the wheel.

Using $T = 2\pi \sqrt{\dfrac{I}{mgh}}$

The restoring torque is generated by the wax.

Re-arranging: $I = \dfrac{T^2 mgh}{4\pi^2}$

Now

$$T = \pi\, s$$
$$m = 1\,g = 10^{-3}\,kg$$
$$g = 10\,m\,s^{-2}$$
$$h = 10\,cm = 10^{-1}\,m$$

$$\therefore\ I = \frac{\pi^2 \times 10^{-3} \times 10 \times 10^{-1}}{4\pi^2}$$

$$I = 2\cdot 5 \times 10^{-4}\,kg\,m^2$$

Practice questions

1   A cylinder of uniform cross sectional area floats to a depth of 0·5 m in a liquid. Show that if it is displaced below the equilibrium position and released it will oscillate with S.H.M. and calculate its period, assuming g = 10 m s^{-2}.

2   A body vibrates with S.H.M. of frequency 100 Hz and amplitude 1·6 mm. Calculate the velocity when it passes through the equilibrium position and the acceleration at maximum displacement.

3   A body moves with S.H.M. When its displacement from the rest position is 3×10^{-2} m, its velocity is $0\cdot 2646y$ m s^{-1}; when its displacement from the rest position is $3\cdot 5 \times 10^{-2}$ m, its velocity is $0\cdot 1936\,\pi$ m s^{-1}. Calculate the amplitude of its motion.

4   A pendulum of length 1 m has a bob of mass 100 gm. Calculate the maximum K.E. of the bob if the amplitude of its motion is 5 cm. Assume g = 10 m s^{-2}.

5   Calculate the radius of a ring which oscillates about an axis through a point on its circumference normal to the plane of the ring with S.H.M. of period π s. Assume g = 10 m s^{-2}.

8 GRAVITATION

Gravitation is a universal effect operating according to the same law between every two particles in the universe.

The fundamental law of gravitation is Newton's **Universal Law of Gravitation**: 'Every **portion** of matter in the universe attracts every other portion with a force which is directly proportional to the product of their masses and inversely proportional to the square of the distance between them'.

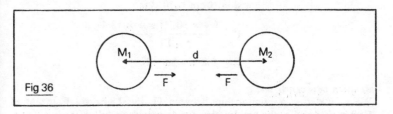

Fig 36

In Fig. 36 we consider two spheres of mass M_1 and M_2 kg whose centres are d m apart.

It can be shown that the mass of a sphere can be considered to be concentrated at its centre.

Newton's Law of Gravitation states:

$$F \propto \frac{M_1 M_2}{d^2}$$

or

$$F = \frac{G M_1 M_2}{d^2}$$

where G is called the **Universal Constant of Gravitation**.
Experimentally G is found to be $6 \cdot 67 \times 10^{-11} \, \text{Nm}^2 \, \text{kg}^{-2}$
Its dimensions are therefore $M^{-1} \, L^3 \, T^{-2}$
Note: Because G is so small, the gravitational forces between most normal objects is very small compared with other forces met in physics. It is therefore known as **a weak force**.

Example
Calculate the force of attraction between two masses of 5 kg and 13·4 kg whose centres are 10 cm apart given $G = 6 \cdot 7 \times 10^{11} \, \text{N m}^2 \, \text{kg}^{-2}$.

Using

$$F = \frac{GM_1M_2}{d^2}$$

$$G = 6 \cdot 7 \times 10^{11} \, \text{N m}^2 \, \text{kg}^{-2}$$

$$M_1 = 5 \, \text{kg}$$

$$M_2 = 13 \cdot 4 \, \text{kg}$$

$$d = 10 \, \text{cm} = 10^{-1} \, \text{m}$$

$$F = \frac{6 \cdot 7 \times 10^{-11} \times 5 \times 13 \cdot 4}{10^{-2}}$$

$$F = 4 \cdot 489 \times 10^{-7} \, \text{N}$$

Relation between G and g

We may derive a relationship between G, the **Universal Constant of Gravitation** and g, the **acceleration due to gravity** as follows:

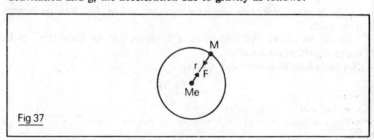

Fig 37

Let the radius of the earth be r m and let its mass be M_e. Let a body of Mass M be situated on the earth's surface and let the force between them be F.

Then

$$F = \frac{GMM_e}{r^2}$$

But

$$F = M \, g$$

Thus

$$g = \frac{GM_e}{r^2}$$

Variation of g

1 Above the surface of the earth g varies inversely with the square of the distance from the centre, as can be seen from the equation above.

2 Below the surface of the earth it can be shown by a mathematical treatment beyond the scope of 'A' level that g varies directly with the distance from the centre.

This is shown in Fig. 38.

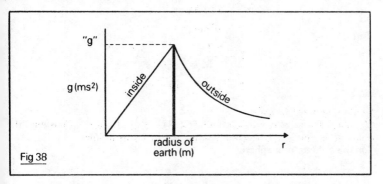

Fig 38

It can be seen in Fig. 38 that g at a certain distance **below** the earth's surface is greater than that at the same distance **above** the earth's surface.

Example

If g on the surface of the earth is $9 \cdot 8 \text{ m s}^{-2}$ and the radius of the earth is 6400 km, calculate the mass of the earth. Assume $G = 6 \cdot 7 \times 10^{-11} \text{ N m}^2 \text{ kg}^{-2}$.

Using

$$g = \frac{GM_e}{r^2}$$

hence

$$M_e = \frac{gr^2}{G}$$

Substituting the values

$$M_e = \frac{9 \cdot 8 \times (6 \cdot 4 \times 10^6)^2}{6 \cdot 7 \times 10^{-11}} = 6 \times 10^{24} \text{ kg}$$

Note: The concept of the 'weight' of the earth has no meaning by definition of weight.

Density of the earth
Assuming the mass of the earth to be M kg and its volume to be V m³ we have:

$$\rho = \frac{M}{V}$$

But

$$M = \frac{gr^2}{G}$$

and

$$V = \frac{4}{3}\pi r^3$$

for the volume of a sphere

$$\rho = \frac{3\,g}{4\pi Gr}$$

Substituting the usual values for g, G and r yields:

$$\rho_{earth} = 5.5 \times 10^3 \text{ kg m}^{-3}$$

Note: We know from experience that the density of the upper layers of the earth's surface have a lower density than this. Hence the material composing the earth must be inhomogeneous.

Inertial and gravitational mass
The mass of a body appearing in $F = ma$ is known as **inertial mass**.

The mass of a body appearing in $F = mMG/r^2$ is known as **gravitational mass**.

Relativity Theory shows the two to be identical.

Synchronous orbits
If a satellite orbits the earth so as to stay over the same place on the earth while the earth rotates, it is said to be in a **synchronous orbit**. The orbit can be calculated as follows:

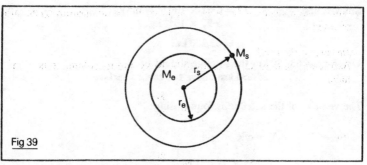

Fig 39

Let the earth have mass M_e kg and radius r_e m.
Let the satellite have mass M_s kg and the radius of its orbit be r_s m.
The force acting F is as follows:

$$F = \frac{G M_e M_s}{r_s^2} \text{ Newton's Law}$$

Note that this is the case because the mass of the earth may be regarded to be concentrated at its centre.

But $\qquad\qquad F = M_s r_s \omega^2 \quad$ Circular Motion

Hence $\qquad M_s r_s \omega^2 = \dfrac{GM_e M_s}{r_s^2}$

$$\therefore \; r_s^3 = \frac{GM_e}{\omega^2}$$

Now, ω, the angular velocity of satellite and planet, can be found from:

$$\omega = \frac{2\pi}{T}$$

T, their period, is 24 hrs = 24 × 3600 = 86,400 s.

$$\therefore \; \omega = \frac{2\pi}{86,400}$$

Hence

$$r_s = \sqrt[3]{\frac{GM_e \, 86,400^2}{4\pi^2}}$$

Substituting
$$G = 6\cdot67 \times 10^{-11} \, N \, m^2 \, kg^{-2}$$
$$M_e = 5\cdot97 \times 10^{24} \, kg$$

$$r_s = 42,200 \, km$$

Using r_e as 63,370 km the satellite should be:
35,830 km above the earth's surface

The velocity of the satellite is found using:

$$V = \omega R = \frac{2\pi}{T} s$$

$$= \frac{2\pi \times 42,200}{86,400} = 3\cdot07 \, m \, s^{-1}$$

Synchronous orbit: 42,200 km, 3·07 km s⁻¹

Weightlessness

When a satellite blasts off from earth on top of its rocket the astronauts are subjected to large force as a result of the strong acceleration. They perceive an increase in weight.

When travelling at constant velocity from the earth to the moon, their weight varies as they move away from the earth. Weightlessness will be perceived when they are at such a distance from earth and moon that the gravitational effects cancel out.

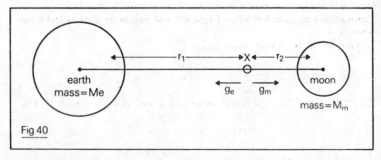

Fig 40

In Fig. 40, consider an astronaut at X. He will be subjected to a gravitational acceleration due to earth, g_e, and one due to the moon, g_m. He will appear weightless when:

$$g_e = g_m$$

i.e.

$$\frac{GM_e}{r_1{}^2} = \frac{GM_m}{r_2{}^2}$$

when

$$\frac{r_2}{r_1} = \sqrt{\frac{M_m}{M_e}}$$

Beyond this point the moon's gravitational acceleration will become more influential.

True weightlessness will again be perceived when the astronaut is in orbit around the moon. His 'centrifugal' acceleration will then balance the gravitational acceleration due to the moon.

Gravitational potential

The gravitational potential at a point is defined as the work done in bringing unit mass from infinity to the point. The unit of measurement is the Joule (J).

Note: This definition is analogous to that of electric potential, except of course that electrical forces may be attractive or repulsive whereas gravitational forces are attractive only.

An expression for the gravitational potential at a point, V

Consider unit mass situated at a distance r m from a body of mass M kg and radius a m.

Force, F, acting on the mass is:

$$F = \frac{GM}{r^2}$$

The work done, dW, by gravity in moving the unit mass a distant dr is thus:

$$dW = F\, dr = \frac{GM}{r^2}\, dr$$

The total work done in bringing the mass from infinity to the surface of the body is:

$$W = \int_{\infty}^{a} \frac{GM}{r^2} \, dr = \left[-\frac{GM}{r} \right]_{\infty}^{a} = -\frac{GM}{a}$$

But, by definition, this is V.

$$\therefore \ V = -\frac{GM}{a}$$

independent of body's mass

Note that it is negative, indicating that the potential at infinity (zero) is greater than it is on the object. This must be the case since gravity has done work in bringing the object to it (the body).

Escape velocity

This is the velocity with which a body must be projected from a planet so as never to return.

An expression for escape velocity
Let a body of mass m kg be projected from a planet of mass M kg.

Work done by body = Mass × Potential

since, mathematically, it will have to travel to infinity to escape the effects of the planet's gravity.

$$\therefore \ \text{Work done} = m \, V$$

This work is done at the expense of the body's kinetic energy. If the velocity of projection is $v \, \text{m s}^{-1}$, then:

$$\text{K.E. of projection} = \tfrac{1}{2} m \, v^2 \, J$$

$$\text{K.E. after escape} = 0$$

$$\therefore \ \text{Loss of K.E.} = \tfrac{1}{2} m \, v^2 \, J$$

$$\therefore \ \tfrac{1}{2} / m \, v^2 = m \, V$$

or

$$v = \sqrt{2 \, V} \ \text{m s}^{-1}$$

independent of body's mass

Example
Calculate the potential on the surface of the earth and the escape velocity for a body situated on the surface of the earth, assuming $G = 6 \cdot 67 \times$

10^{-11} N m² kg⁻², mass of earth $= 5.97 \times 10^{24}$ kg, radius of earth $= 6.37 \times 10^6$ m.

Using

$$V = - \frac{GM}{a}$$

Substituting

$$V = \frac{-6.67 \times 10^{-11} \times 5.97 \times 10^{24}}{6.37 \times 10^6}$$

$$V = -6.25 \times 10^7 \text{ J}$$

Using

$$v = \sqrt{2V}$$
$$= \sqrt{2 \times 6.25 \times 10^7}$$
$$= \sqrt{12.5 \times 10^7}$$
$$v = 1.12 \times 10^4 \text{ m s}^{-1}$$

A simplified expression for escape velocity

A simplified expression for escape velocity for a body on a planet of known radius, having a known acceleration due to gravity on its surface, may be derived as follows:

$$V = (-)\frac{GM}{a}$$

But

$$\frac{GM}{a} = a\,g$$

$$\therefore\ V = a\,g$$

$$\therefore\ v = \sqrt{2V} = \sqrt{2\,g\,a}$$

$$v = \sqrt{2\,g\,a} \text{ m s}^{-1}$$

Determination of the acceleration due to gravity by free fall

A ball bearing is held by an electromagnet. It is released and simultaneously a centisecond timer is started. After falling a known distance it hits a metal plate which stops the timer. The experiment is repeated at least five more times to give a set of distances and corresponding times.

Theory
Using

$$s = ut + \tfrac{1}{2}at^2$$

For a body falling from rest under gravity:

$$s = \frac{g}{2}t^2$$

A graph of s against t^2 will have a gradient g/2. Hence plot a graph and double measured gradient.

Practice questions

1   The gravitational force between two masses of 2·6 kg and 8·9 kg is 1·382 × 10^{-8} N. Calculate the distance between their centres if G = 6·7 × 10^{-11} N m^2 kg^{-2}.

2   Calculate a value for g on the surface of the sun, assuming it to be a sphere of radius 6·98 × 10^8 m and mass 1·99 × 10^{30} kg. Take G = 6·67 × 10^{-11} N m^2 kg^{-2}.

3   Assuming that Mars is a sphere of radius 3·40 × 10^6 m and mass 6·42 × 10^{23} kg, find a value for the potential on its surface and the escape velocity for a body on its surface. G = 6·67 × 10^{-11} N m^2 kg^{-2}.

9 VISCOSITY

Liquids resist stirring to some degree or other. This is due to their viscosity which is, in effect, 'liquid friction'.

Coefficient of viscosity

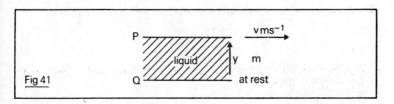

Fig 41

A plane P of area A m^2 moves parallel to a fixed plane Q with a velocity v m s^{-1}. The separation between them is y m and the space is filled with liquid.

Providing v is not too great, each layer of the liquid moves steadily as a layer in **laminar** or **streamline** flow. (Otherwise the motion is known as **turbulent**.) The force, F, with which P is moved along (at constant velocity) balances the viscous forces.

The **coefficient of viscosity**, η is defined by:

$$F = \eta A \frac{dv}{dy}$$

Thus

$$\eta = \frac{F}{A} \frac{dy}{dv}$$

Units: $N\,m^{-2}\,s$ Dimensions: $ML^{-1}\,T^{-1}$

It is a function of temperature, decreasing as temperature increases.

Example

If the coefficient of viscosity (at a particular temperature) of water is $1 \cdot 3 \times 10^{-3}\,N\,m^{-2}\,s$, calculate the frictional force on a plate of area $10^{-3}\,m^2$ moving at a velocity of $2 \times 10^{-2}\,m\,s^{-1}$ with respect to a fixed plate and separated from it by a layer of water $2 \times 10^{-3}\,m$ thick.

Note: In most of these problems we do not know v as a function of y. However, the separation is usually so small that we replace dv/dy by v/y.

Hence

$$F = \eta A \frac{v}{y}$$

$$= 1 \cdot 3 \times 10^{-3} \times 10^{-3} \times \frac{2 \times 10^{-2}}{2 \times 10^{-3}} = \mathbf{1 \cdot 3 \times 10^{-5} \, N}$$

Poiseuille's Formula

The volume of a liquid dV/dt issuing every second from a pipe is given by

$$\frac{dV}{dt} = \frac{\pi p a^4}{8 \eta L}$$

p = pressure difference across pipe
a = radius of pipe
η = coeff. of viscosity
L = length of pipe

The formula is often used experimentally to find the Coefficient of Viscosity of a liquid.

Example

A liquid flows through a horizontal tube of length 0·25 m and of internal radius 5×10^{-4} m, across which the pressure difference is 1000 N m^{-2}. If the rate of flow is 6×10^{-11} m^3 s^{-1}, calculate the coefficient of viscosity of the liquid.

Since

$$\frac{dV}{dt} = \frac{\pi p a^4}{8 \eta L}$$

$$\eta = \frac{\pi p a^4}{8 L (dV/dt)}$$

Substituting:

$$\eta = \frac{\pi \times 1000 \times (5 \times 10^{-4})^4}{8 \times 0 \cdot 25 \times 6 \times 10^{-11}}$$

$$= \mathbf{1 \cdot 636 \, N \, m^{-2} \, s}$$

Terminal velocity

When a small object is dropped into a viscous liquid, we find that the viscous retarding force acting on the object is proportional to its velocity.

The equation describing an acceleration (or force) proportional to velocity is:

$$m\frac{d^2x}{dt^2} = k\frac{dx}{dt}$$

A situation is reached when:

upward force = downward force

i.e. viscous retarding force + upthrust = weight

At this point, the velocity of the object reaches a steady maximum known as the **terminal velocity**.

Fig. 42 shows how the terminal velocity is reached.

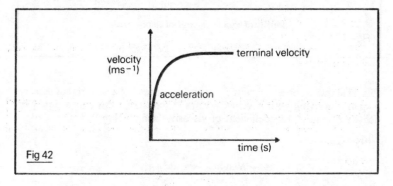

Fig 42

Notes:

1 In the case of the free-fall experiment for the determination of g, both the upthrust and the viscous retarding force are negligible compared with the weight.

2 In the case of raindrops, however, the viscous retarding force is appreciable because of their velocity. This viscous force (and the upthrust which is still negligible) balances the weight so that raindrops do reach their terminal velocity. If this were not the case, their velocity on reaching the earth's surface would be so great as to cause damage.

Stokes' Law

If a sphere of radius a is dropped into a viscous liquid of coefficient of viscosity η and its velocity at any instant is v then the viscous retarding force, F is given by:

$$F = 6\pi\eta av \text{ N} \quad \text{Stokes' Law}$$

A sphere falling with terminal velocity through a medium

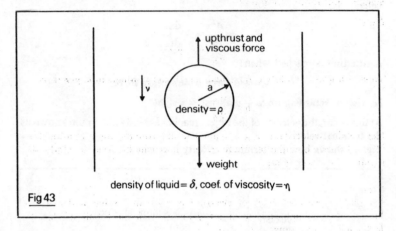

Fig 43

Fig. 43 shows a sphere of radius a m and composed of a material of density ρ kg m^{-3} moving with a velocity v m s^{-1} (constant) through a liquid of density δ kg m^{-3}, its coefficient of viscosity being η N m^{-2} s.

Now

$$\text{Weight} = \frac{4}{3}\pi a^3 \rho\, g$$

$$\text{Upthrust} = \frac{4}{3}\pi a^3 \delta\, g$$

$$\text{Viscous force} = 6\pi\eta a v$$

Hence

$$\frac{4}{3}\pi a^3 \rho\, g = \frac{4}{3}\pi a^3 \delta\, g + 6\pi\eta a v$$

or

$$\eta = \frac{2}{9}\frac{a^2 g}{v}(\rho - \delta)\,\text{N m}^{-2}\,\text{s}$$

This formula is often used to determine experimentally the coefficient of viscosity of a liquid. Spheres of different radii are timed falling between known distances in the liquid. Thus corresponding values (at least six) of velocity and radius are found.

Now

$$v = \frac{2g}{9\eta}(\rho - \delta)a^2$$

Thus, if a graph is plotted on v against a^2, the gradient m is given by

$$m = \frac{2g}{9\eta}(\rho - \delta)$$

Hence m is measured on the graph and η found from the above relationship.
Note: Experiment ignores 'edge' effects and can be used to study the variation of η with temp.

Example
A liquid whose coefficient of viscosity is $2 \cdot 2\,N\,m^{-2}\,s$ has a density of $960\,g\,m^{-3}$. Calculate the terminal velocity of a steel ball of density $7800\,g\,m^{-3}$ if its radius is $2 \times 10^{-3}\,m$.

Use

$$V = \frac{2g}{9\eta}a^2(\rho - \delta)$$

Substituting

$$V = 2 \times 10 \times 4 \times 10^{-6} \times 6840$$
$$V = 2 \cdot 764 \times 10^{-2}\,m\,s^{-1}$$

Practice questions

1 A flat plate of area $0 \cdot 1\,m^2$ is placed on a flat surface and is separated from it by an oil film of thickness $10^{-5}\,m$ which has a coefficient of viscosity of $1 \cdot 0\,N\,m^{-2}\,s$. Calculate the force needed to cause the plate to move with a constant velocity of $10^{-3}\,m\,s^{-1}$.

2 What would be the rate of flow in the previous example if the pressure difference were doubled, the radius doubled and the length multiplied by a factor of 32?

3 Calculate the viscous retarding force acting on a sphere of radius $2 \times 10^{-2}\,m$ falling at a velocity of $0 \cdot 15\,m\,s^{-1}$ through a medium of coefficient of viscosity $1 \cdot 5\,N\,m^{-2}\,s$.

4   Sand particles are sprinkled on to the surface of the water in a beaker filled to a depth of 0·15 m and the grains sink immediately with their terminal velocity. Calculate the time taken for grains of radius 5 × 10^{-5} m to reach the bottom, assuming $g = 10$ m s^{-2}, density of sand = 2200 kg m^{-3}, coefffficient of viscosity of water = $1·1 \times 10^{-3}$ N m^{-2}.

10 SURFACE TENSION

Observations suggest that the surface of a liquid acts like an elastic skin covering the liquid i.e. the liquid surface is in a state of **'tension'**.

The effect is due to asymmetric intermolecular forces. It is known as the **surface tension effect**.

The surface tension, T, of a liquid is defined as the force per unit length acting in the surface of the liquid at right angles to one side of a line drawn in the surface of that liquid.

Units: $N\,m^{-1}$ **Dimensions:** $M\,T^{-2}$

Surface tension is a function of temperature, decreasing as temperature rises. For water at 293 K it is $7\cdot3 \times 10^{-2}\,N\,m^{-1}$.

Surface tension phenomena:

1   Certain insects can walk on water.
2   Water drops can remain suspended from a tap before falling.
3   Dry, clean needles may be made to float on the surface of water.

Pressure difference across a curved liquid surface

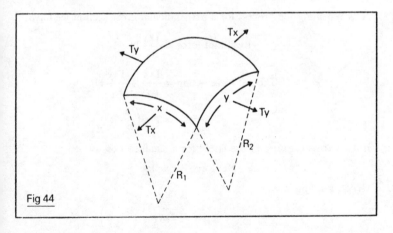

Fig 44

Consider a curved liquid surface as shown in Fig. 44 (imagine a telephone kiosk roof).

Let the arc lengths be x and y m and the radii of curvature be R_1 and R_2 m and let the surface tension of the liquid be $T\,N\,m^{-1}$.

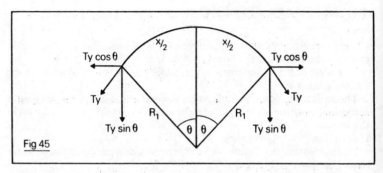

Fig 45

Consider a cross section as shown in Fig. 45. Resolving the surface tension forces we have:

Horizontally: 0

Vertically: $2 \, Ty \, \sin \theta$ N

Now for a small portion of surface θ is small and $\sin \theta \approx \theta$.

$\therefore$ Resultant force $= 2Ty\theta = 2Ty\dfrac{x/2}{R_1} = \dfrac{Tyx}{R_1}$ N

Now repeating the process for a perpendicular cross section we have:

$$\text{Resultant force} = \frac{Txy}{R_2} \text{ N}$$

$$\text{The net force acting} = \frac{Tyx}{R_1} + \frac{Txy}{R_2} \text{ N}$$

$$= Txy\left(\frac{1}{R_1} + \frac{1}{R_2}\right) \text{N}$$

If ρ = excess pressure in the liquid for a small section xy

$$\text{Net force} = pxy \text{ Pa}$$

Hence, for equilibrium:

$$pxy = Txy\left(\frac{1}{R_1} + \frac{1}{R_2}\right)$$

or

$$P = T\left(\frac{1}{R_1} + \frac{1}{R_2}\right) Pa$$

P is the excess pressure, in Newtons, in a drop of liquid due to surface tension.

In the case of a film there are 2 surfaces, hence:

$$p = 2T\left(\frac{1}{R_1} + \frac{1}{R_2}\right) Pa$$

For a spherical drop of liquid in air (or an air bubble in liquid) we have equal radii of curvature:

$$\therefore \ p = \frac{2T}{R} Pa$$

In the case of a spherical film i.e. a soap bubble in air then

$$p = \frac{4T}{R} Pa$$

Note: An interesting case arises from this last equation. It can be seen that the smaller the radius of a soap bubble, the greater is the excess pressure. Thus if two soap bubbles of differing radii are blown on a tapped U tube and the taps opened, contrary to common sense expectations, the smaller bubble would collapse at the expense of the larger bubble.

Example
A soap bubble has a radius of 10^{-3} m. Assuming atmospheric pressure $= 10^5$ Pa and the surface tension of soap solution at the prevailing temperature $= 2 \cdot 8 \times 10^{-2} \ N \ m^{-1}$, calculate the pressure inside it.

Using

$$p = \frac{4T}{R} = \frac{4 \times 2 \cdot 8 \times 10^{-2}}{10^{-3}}$$

$$= 112 \ Pa$$

$$\therefore \ Total \ pressure = 10^5 + 112$$
$$= 10^5 + 0 \cdot 00112 \times 10^5$$
$$= 1 \cdot 00112 \times 10^5 \ Pa$$

Angle of contact

The form of a liquid surface in the vicinity of a solid depends upon the forces exerted on each liquid molecule by the liquid (cohesive forces) and solid molecules (adhesive forces) in its vicinity.

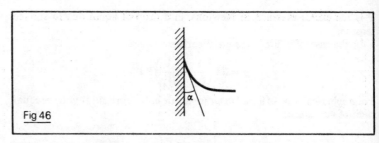

Fig 46

In Fig. 46, the adhesive forces are very large and the liquid wets the solid e.g. pure water on clean glass.

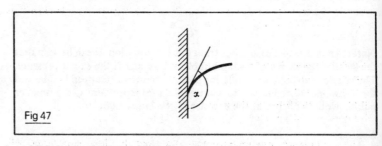

Fig 47

In Fig. 47 the cohesive forces are very large and the liquid does not wet the solid e.g. mercury on clean glass.

The **Angle of Contact**, α, between a solid and a liquid is that angle, measured within the liquid, between the solid surface and the tangent plane to the liquid surface at the point of intersection.

The value of α for water/glass varies from 0 to about 8° and for mercury/glass it is about 130°.

Capillarity

When a capillary tube is immersed in water and then placed vertically with one end in the liquid, the water rises in the tube above the external free surface. If the procedure is repeated with mercury, the mercury falls below the level outside the tube.

The effect is known as **capillarity** and it can be understood in terms of the vertical components of the surface tension forces acting at the liquid/capillary intersection (see Figs 46 & 47).

To calculate the rise of a liquid in a capillary tube

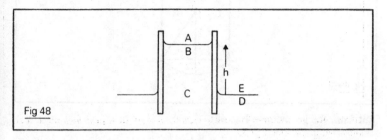

Fig 48

Fig. 48 shows a cross section of a capillary tube in a vertical plane with its bottom end in a liquid.

Let:

Surface tension of liquid $= T \, N \, m^{-1}$
Angle of contact $= \theta°$
Density of liquid $= \rho \, kg \, m^{-3}$
Radius of curvature of liquid surface $= R \, m$
Radius of capillary tube $= r \, m$

Note: In practice, because the bore is very small, the plane of the liquid will, in fact, be hemispherical.

Then: Pressure at **A** = Atmospheric

    Pressure at **B** = Atmospheric $- 2T/R$

    Pressure at **C** = Atmospheric $- 2T/R + \rho gh$

    Pressure at **D** = Atmospheric $- 2T/R + \rho gh$

    Pressure at **E** = Atmospheric

But Pressure at **D** = Pressure at **E**

$$\therefore \text{ Atmospheric} - \frac{2T}{R} + \rho gh = \text{Atmospheric}$$

$$\therefore h = \frac{2T}{\rho gR}$$

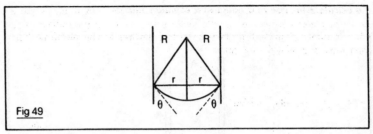

Fig 49

But from the geometry – Fig. 49:

$$r = R \cos \theta$$

$$\therefore \frac{1}{R} = \frac{\cos \theta}{r}$$

Hence:

$$h = \frac{2T \cos \theta}{\rho g r} \text{ m}$$

From this if $\theta < 90$, $\cos \theta$ is +ve $\therefore$ h is positive
From this if $\theta > 90$, $\cos \theta$ is −ve $\therefore$ h is negative
This method can be used experimentally for determining the surface tension of a liquid.

Example
A capillary tube of radius 2×10^{-4} is placed vertically inside water of surface tension $7 \cdot 3 \times 10^{-2} \text{ N m}^{-1}$ and zero angle of contact. Calculate the height to which the water rises. Assume $g = 10 \text{ m s}^{-2}$ and density of water $= 1000 \text{ kg m}^{-3}$.
Using

$$h = \frac{2T \cos \theta}{\rho g r}$$

Substituting:

$$h = \frac{2 \times 7 \cdot 3 \times 10^{-2} \times 1}{1000 \times 10 \times 2 \times 10^{-4}} = 7 \cdot 3 \times 10^{-2} \text{ m}$$

Measurement of surface tension by Jaeger's method

This allows the surface tension and also its variation with temperature to be determined.

Basically Jaeger's method consists of measuring the maximum pressure required to produce an air bubble at the end of a vertical capillary tube which is immersed in the liquid whose surface tension is to be measured. The apparatus is shown in Fig. 50.

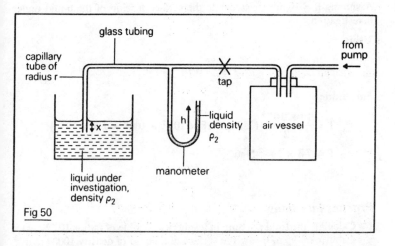

Fig 50

The tap is adjusted until bubbles form at the end of the capillary at the rate of one every few seconds. The maximum pressure difference of the liquid levels in the manometer is noted and hence the maximum pressure in the bubbles before they break away from the capillary is found.

Pressure inside bubble, p_1 = atmospheric + $\rho_2 gh$

Pressure outside bubble p_2 = atmospheric + $\rho_1 gx$

$\therefore$ Excess pressure = $p_1 - p_2 = g(\rho_2 h - \rho_1 x)$

But excess pressure = $\dfrac{2T}{r}$

$$\therefore T = \frac{gr}{2}(\rho_2 h - \rho_1 x) \, \text{N m}^{-1}$$

Note: The variation of T with temp can be noted. However we ignore the simultaneous variation in ρ_1 as temperature changes. This can be overcome, when accurate determination is required, in one of two ways:

1 using the relationship between density and temperature
2 using a modified experiment outside the scope of 'A' level.

Example

In Jaeger's experiment the following values were obtained:

$r = 10^{-4}$ m, $\qquad \rho_1 = 1000$ kg m^{-3}, $\quad \rho_2 = 750$ kg m^{-3}
$x = 2\cdot5 \times 10^{-2}$ m, $\quad h = 2\cdot28 \times 10^{-1}$ m

Assuming $g = 10$ m s^{-2}, calculate the surface tension of the liquid under investigation.

Using

$$T = \frac{gr}{2}(\rho_2 h - \rho_1 x)$$

Substituting:

$$T = \frac{10 \times 10^{-4}}{2}(750 \times 2\cdot28 \times 10^{-1} - 1000 \times 2\cdot5 \times 10^{-2})$$

$$T = 7\cdot8 \times 10^{-2}\,N\,m^{-1}$$

Practice questions

1 Calculate the total pressure in a spherical bubble of radius 10^{-3} m at a distance of 0·5 m below the surface of a liquid of density 1000 kg m^{-3} whose surface tension is $7\cdot3 \times 10^{-2}$ N m^{-1}, assuming that the atmospheric pressure is 10^5 Pa and $g = 10$ m s^{-2}.

2 A glass capillary tube is placed vertically with one end dipping into a liquid of density 900 kg m^{-3} and surface tension $2\cdot5 \times 10^{-2}$ N m^{-1}, the angle of contact being 30°. If the liquid rises $4\cdot5 \times 10^{-2}$ m, calculate the radius of the tube. Assume $g = 10$ m s^{-2}.

3 In a Jaeger's experiment the following values were obtained:
   diameter of capillary $= 3\cdot5 \times 10^{-4}$ m
   density of manometer liquid $= 800$ kg m^{-3}
   density of liquid under investigation $= 1000$ kg m^{-3}
   length of capillary immersed in liquid $= 3\cdot2 \times 10^{-2}$ m
   difference between manometer levels $= 10^{-1}$ m.
   Assuming $g = 10$ m s^{-2}, calculate the value of the surface tension of the liquid.

11 ELASTICITY

Behaviour of bodies under stress, types of deformation

The behaviour of a body subjected to an externally applied force depends upon the nature of the material, the magnitude of the force and the previous history of the specimen.

Solids which can be hammered into flat sheets are said to be **malleable**; those which can be drawn into wires are said to be **ductile**.

A solid which is able to recover its original dimensions when an applied force is removed from it is said to be **elastic** and exhibits the property of elasticity. One which does not recover but remains permanently deformed is said to be **plastic** and exhibits the property of **plasticity**.

Hooke's Law

Elastic materials behave as such unless sufficient force is applied to them to cause them to be deformed beyond a critical limit known as the **elastic limit**; beyond this elastic behaviour ceases.

Hookes' Law, which applies to wires and springs, states 'Provided the elastic limit is not exceeded, the extension is directly proportional to the applied force'.

$$F = kx$$

F = force in Newtons
x = extension in metres

k is the elastic constant for the wire or spring and has units of $N\,m^{-1}$. A graph of force against extension is therefore a straight line through the origin, of gradient k.

Acceleration due to gravity using an oscillating spring

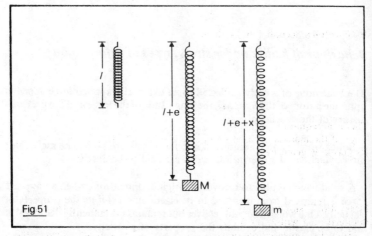

Fig 51

As shown in Fig. 51, imagine we have a spring of natural length l. If a mass M is hung on its end the spring will stretch by an amount e (less than the elastic limit).

By Hooke's Law, the force and extension are related by:

$$F = ke$$

This is the downward force on the spring. By Newton's Third Law, this is also an upward force on the mass.
The downward force on the mass is its weight.

$$F = Mg$$

Since the mass is at rest, upward and downward forces are equal:

$$\therefore \ Mg = ke$$

If the mass is now pulled down a further distance x:

The net upward force now acting is

$$F_{UP} = k(e + x)$$

The downward force is still

$$F_D = Mg = ke$$

$$\therefore \ \text{Resultant force} = k(e + x) - ke = kx$$

This acts upward to decrease x — **it is a restoring force**

$$\therefore \ F = -kx$$

By Newton's Second Law we have

$$M\frac{d^2x}{dt^2} = kx$$

or

$$\frac{d^2x}{dt^2} = -\frac{k}{m}x$$

This is an equation of S.H.M.

Thus, if the mass is released it will oscillate with S.H.M.

Its period is given by $T = 2\pi/\omega$

$$\therefore \ T = 2\pi\sqrt{\frac{M}{k}}$$

Note: For a 'sloppy' spring, k is small and therefore T is large. This can be demonstrated with a 'slinky'.

Using the previous relationships, we can find g without knowing the elastic constant of the spring, provided that the static extension is known.

Now

$$T = 2\pi\sqrt{\frac{m}{k}}$$

But $Mg = ke$

$$\therefore \ T = 2\pi\sqrt{\frac{e}{g}}$$

or

$$T^2 = \frac{4\pi^2}{g}\rho$$

Thus, experimentally, the spring is loaded to particular static extension which is noted. It is then pulled down an extra distance x (x < e to avoid overshoot) and 20 oscillations timed and the period found. The process is repeated at least five more times and a graph plotted of T^2 against e. Its gradient, m, is measured.

$$m = \frac{4\pi^2}{g} \text{ and hence g}$$

Note: This simple treatment neglects the mass of the spring. If this is not negligible, a full treatment shows that the mass used should be replaced by an effective mass M_e given by:

$$M_e = M + \frac{\text{Mass of spring}}{3}$$

Example

If a mass of 0.2 kg is suspended from a vertical spring of negligible mass, the extension caused is 4×10^{-2} m. The mass is replaced by one of 0.5 kg and, after equilibrium has been reached, it is pulled down and released. Assuming $g = 10 \text{ m s}^{-2}$, calculate the period of the motion.

Use

$$T = 2\pi \sqrt{\frac{M}{k}}$$

To find k use

$$F = ke$$

with

$$F = mg = 0.2 \times 10 = 2 \text{ N}$$

$$e = 4 \times 10^{-2} \text{ m}$$

$$\therefore k = \frac{F}{e} = \frac{2 \text{ N}}{4 \times 10^{-2} \text{ m}} = 50 \text{ N m}^{-1}$$

Now

$$M = 0.5 \text{ kg}$$

$$\therefore T = 2\pi \sqrt{\frac{0.5}{50}} = 2\pi \sqrt{10^{-2}} = \mathbf{0.2\pi}$$

$$\mathbf{T = 0.2\pi \text{ s}}$$

Tensile stress, tensile strain and Young's modulus

Tensile stress is defined as tensile force per unit area. (By tensile we mean axial or longitudinal.)
Units: N m^{-2} **or** Pa
Dimensions: $M L^{-1} T^{-2}$

Tensile strain is defined as the fractional increase in length. Since it is a ratio it has neither units nor dimensions.

Young's Modulus is defined as the ratio of tensile stress to tensile strain. (The elastic limit not having been exceeded.)

Its units and dimensions are those of stress i.e. N m^{-2} (or Pa) and M L^{-1} T^{-2}

Formula for Young's Modulus

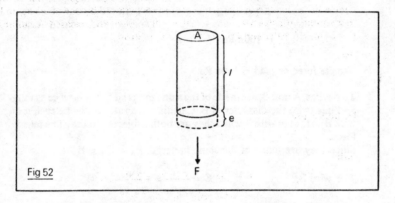

Fig 52

Consider the cylinder shown in Fig. 52.

Its cross sectional area is A and its natural length is l. An extension e is caused by a tensile force F.

$$\text{Tensile stress} = \frac{F}{A}\,\text{Pa}$$

$$\text{Tensile strain} = \frac{e}{l}$$

$$\therefore E = \text{Young's Modulus} = \frac{F/A}{e/l}$$

$$E = \frac{Fl}{Ae}\,\text{Pa}$$

Typical values of E: Mild steel $2 \cdot 0 \times 10^{11}$ Pa
Copper $1 \cdot 2 \times 10^{11}$ Pa
Brass $1 \cdot 0 \times 10^{11}$ Pa

Examples

1   Find the maximum load to which a brass wire of radius 10^{-3} m may be subjected if the strain must not exceed 10^{-3}, given Young's Modulus for brass $= 1 \cdot 0 \times 10^{11}$ Pa.

Using $$E = \frac{\text{tensile stress}}{\text{tensile strain}} = \frac{\text{tensile force}}{\text{area} \times \text{tensile strain}}$$

$\therefore$ Tensile force $= E \times$ area $\times$ tensile strain

$$= 1 \cdot 0 \times 10^{11} \times \pi \times 10^{-6} \times 10^{-3}$$

$$= 100\pi \text{ Pa}$$

Tensile force or load $= 100\pi$ Pa

2   Two wires, A and B, are made of the same material but A is twice as long and has twice the diameter of B. Calculate the ratio of the extension of wire B as that of wire A when they are both subjected to the same tensile force.

Since they are made of the same material, $E_A = E_B = E$

For wire B, $$E = \frac{Fl_B}{A_B e_B} = \frac{Fl_B}{\pi r_B{}^2 e_B}$$

where $F =$ tensile force, $l_B =$ length of B
   $r_B =$ radius of B, $e_B =$ extension of wire B

Now for wire A, $$E = \frac{Fl_A}{\pi r_A{}^2 e_A}$$

Subscript A denoting wire A

Hence equating $$\frac{Fl_B}{\pi r_B{}^2 e_B} = \frac{Fl_A}{\pi r_A{}^2 e_A}$$

Hence

$$\frac{e_B}{e_A} = \frac{l_B r_A{}^2}{l_A r_B{}^2}$$

But, we are told:

$$l_A = 2l_B$$
$$r_A = 2r_B$$

$$\therefore \frac{e_B}{e_A} = \frac{l_B(2r_B)^2}{2l_Br_B{}^2}$$

$$= \frac{4l_Br_B{}^2}{2l_Br_B{}^2} = 2$$

$$\therefore e_B : l_A = 2 : 1$$

i.e. **Extension of B : Extension of A = 2 : 1**

The behaviour of a wire when subjected to Tensile Stress
If a varying tensile stress is applied to wire and the corresponding strain noted, we obtain a graph as shown in Fig. 53.

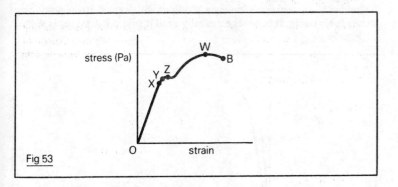

Fig 53

Along OX: stress directly proportional to strain.

Along XY: for this **very** small portion of the graph stress and strain are no longer exactly directly proportional, although we still have elastic behaviour.

X is known as the **proportional limit**

Y is known as the **elastic limit**

Along YZ, if the stress were to be removed, the wire would still have a permanent extension or 'permanent set'.

Z is known as the **yield point**.

Beyond Z the behaviour is plastic and, after the small 'knee' the strain grows rapidly until, at W, the maximum breaking stress is reached. Beyond this point, the strain increases even with reduced stress until the breaking point is reached.

Behaviour of particular materials

Lead, copper and **wrought iron** extend considerably before they break – they are ductile.

Glass and **high carbon steels** break soon after becoming plastic – they are brittle.

Brass and **bronze**, although having an elastic limit, do not appear to have a yield point.

Rubber has a very small Young's Modulus with a small stress producing measurable strain. It behaves elastically until it snaps. Behaviour is shown in Fig. 54.

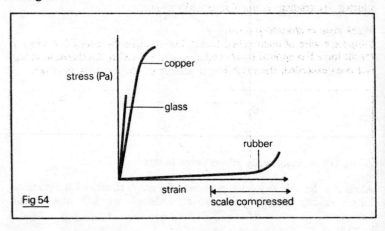

Fig 54

The reasons for this difference in behaviour are basically as follows:
In ordered solids the atoms are arranged in a pattern known as a **crystal lattice**. However the lattice is not perfect – at all temperatures except OK it contains sites with missing atoms known as **vacancies**; it also contains impurity atoms and flaws in the lattice regularity known as **dislocations**. In the case of elastic behaviour, the atomic layers 'slide' over each other so that relative motion occurs in the direction of the stress. When the stress is removed they 'slide' back.

However, when the elastic limit is exceeded the large tensile stress causes movement of the whole crystal plane along the dislocations. Any further discussion is outside the scope of 'A' level since it involves thermodynamics and probability.

In the case of rubber not subjected to stress the molecules are curled up. Application of stress causes their length to increase by straightening them out. The process is identical to the way in which hair may be stretched extensively by straightening out its keratin molecules – this is done by applying stress to wet hair.

Experimental determination of Young's Modulus for a wire

Two wires are suspended from a rigid support. Both are initially stressed. One wire has a vernier scale attached to it and is connected to the other wire by means of a cradle. The other wire is subjected to varying loads and its extension is measured with a micrometer. The initial length of wire is found and also its area. Corresponding values of stress and strain within the proportional region are obtained and a stress against strain graph is plotted. Its gradient is the Young's Modulus.

Work done in stretching a wire

Suppose a wire of unstretched length l is stretched by a length e when a tensile force F is applied to one end of area A. Providing the elastic limit has not been exceeded, the extension is directly proportional to the load.

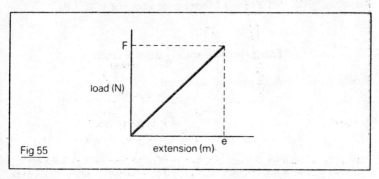

Fig 55

As can be seen from Fig. 55, the average value of the force is F/2.

Now work = Force × Distance

($\cos \theta = 1$ here)

Here, work = Average Force × Distance

$\therefore$ Work done $= \dfrac{F}{2} \times e$

This is stored in the wire in the form of P.E.

$$\text{Work done} = \frac{Fe}{2}$$

Note: We arrive at the same formula even if a more rigorous method is used.

Further, the work done (or energy stored) per unit volume is given by (Vol of cylinder $= Al$)

$$\frac{Fe/2}{Al}$$

$$= \frac{1}{2} \frac{F}{A} \frac{e}{l}$$

$$= \frac{1}{2} \times \text{stress} \times \text{strain}$$

Example

A steel wire of length 2 m and radius 5×10^{-4}'m is stretched by 4×10^{-3} m. Calculate the energy stored per unit volume as a result of this stretching, given that Young's Modulus for steel $= 2 \times 10^{11}$ Pa.

Use

$$\text{Energy per unit vol} = \frac{l}{2} \text{stress} \times \text{strain}$$

Firstly the tensile force must be calculated.

We use

$$E = \frac{Fl}{Ae}$$

$$\therefore F = \frac{EAe}{l}$$

Substituting the values

$$F = \frac{2 \times 10^{11} \times \pi \times 25 \times 10^{-8} \times 4 \times 10^{-3}}{2}$$

$$F = 100\pi \text{ N}$$

Hence

$$\text{Stress} = \frac{F}{A} = \frac{100\pi}{\pi \times 25 \times 10^{-8}} = 4 \times 10^8 \text{ Pa}$$

$$\text{Strain} = \frac{e}{l} = \frac{4 \times 10^{-3}}{2} = 2 \times 10^{-3}$$

$\therefore$ Energy stored per unit vol $= \dfrac{l}{2} \times 4 \times 10^8 \times 2 \times 10^{-3}$

$$= 4 \times 10^5 \text{ J}$$

Energy stored per unit vol $= 4 \times 10^5 \text{ J}$

Force in a bar due to contraction or expansion

If a bar is heated it expands and then contracts when it cools. (The opposite occurs if it is cooled then warmed.) If, however, it is prevented from contracting as it cools (or from expanding as it warms) considerable stresses are developed.

Consider a bar of length l, cross sectional area A and linear expansivity α. Suppose it is heated by t K
The expansion will be $l\alpha\,t$ m

If it is now cooled by t K and prevented from contracting the 'virtual' contraction is $l\alpha\,t$.

Now

$$E = \frac{Fl}{Ae} = \frac{Fl}{Al\alpha t} = \frac{F}{A\alpha t}$$

$\therefore$ **Thermal stress developed $= E\,\alpha\,t$**

Example
Calculate the thermal stress developed in a steel rod of area 10^{-4} m^2 which is rigidly clamped whilst being cooled by 100 K. Assume Young's Modulus for steel $= 2 \times 10^{11}$ Pa and the linear expansivity of steel $= 12 \times 10^{-6}$ K^{-1}.
Use Thermal stress $= E\,\alpha\,t$
Substituting:

$$\text{Thermal stress} = 2 \times 10^{11} \times 12 \times 10^{-6} \times 100$$
$$= 2{\cdot}4 \times 10^8 \text{ Pa}$$
Thermal stress $= 2{\cdot}4 \times 10^8$ Pa

Practice questions

1 Assuming that Hooke's Law applies and $g = 10\,\mathrm{m\,s^{-2}}$, calculate the period of oscillation of a light spring loaded with a mass of $0\cdot2\,\mathrm{kg}$ if a mass of $10^{-3}\,\mathrm{kg}$ causes an extension of $10^{-3}\,\mathrm{m}$.

2 A mass of $4\,\mathrm{kg}$ is hung from a vertical wire $4\,\mathrm{m}$ long, radius $10^{-3}\,\mathrm{m}$, causing an extension of $4\cdot8 \times 10^{-4}\,\mathrm{m}$. Calculate the tensile stress, tensile strain and Young's Modulus for the material of the wire, assuming $g = 10\,\mathrm{m\,s^{-2}}$.

3 What stress would cause a wire to increase in length by 1% if Young's Modulus for the wire is $1\cdot3 \times 10^{11}\,\mathrm{Pa}$?

4 A steel wire of length $2\,\mathrm{m}$ and radius $10^{-3}\,\mathrm{m}$ is subjected to a tensile force of 40N. If Young's Modulus for steel is $2 \times 10^{11}\,\mathrm{Pa}$, calculate the strain energy stored per unit volume.

5 Calculate the thermal stress developed in a steel rod of radius $5 \times 10^{-3}\,\mathrm{m}$ which is rigidly clamped and heated by 20 K. Assume Young's Modulus for steel $= 2 \times 10^{11}\,\mathrm{Pa}$ and linear expansivity of steel $= 1\cdot1 \times 10^{-5}\,\mathrm{K^{-1}}$.

12 OMNIBUS WORKED EXAMPLES

1   A bullet of mass 0·006 kg is fired from a gun which recoils with a velocity of 3 m s⁻¹. The bullet is retarded, due to air resistance, at the rate of 2 m s⁻². After 50 s its velocity is 800 m s⁻¹. Find the mass of the gun, naming any laws or equations which you use.

Firstly, we find the velocity of the bullet when it left the gun – its muzzle velocity – using the equations of motion. Secondly, we find the mass of the gun using the principle of Conservation of Momentum.

(i) Let u = muzzle velocity of bullet
then v = (its velocity after 50 s) is 800 m s⁻¹
t = (time between u and v) is 50 s
a = (retardation of bullet) is 2 m s⁻²
Now $v = u + at$
thus $800 = u - 2 \times 50$
$$u = 900 \text{ m s}^{-1}$$

(ii) Mass of gun × recoil velocity = Mass of bullet × muzzle velocity

$$m \times 3 = 0.006 \times 900$$

$$\therefore \ m = \frac{0.006 \times 900}{3} = 1.8 \text{ kg}$$

Mass of gun = 1·8 kg

2   A wheel is spinning steadily at 10 revs per second about an axis through its centre. If the moment of inertia about the axis is 0·2 kg m², calculate the kinetic energy and angular momentum of the wheel. What is the torque of the constant couple which brings the wheel to rest in 50 revs?

Use

$$\text{K.E.} = \tfrac{1}{2} I \omega^2$$

where I = Moment of Inertia
ω = Angular velocity (in r s⁻¹)
$$I = 0.2 \text{ kg m}^2$$

$$10 \text{ rev s}^{-1} = 20\pi \text{ r s}^{-1} = \omega$$

$$\therefore \ \text{K.E.} = \frac{1}{2} \times 0.2 \times (20\pi)^2$$

$$\text{K.E.} = 394\cdot8 \text{ J}$$

Use Angular Momentum $= I \omega$
$$= 0.2 \times 20\pi$$

$$= 4\pi$$

Angular Momentum $= 4\pi \text{ kg m}^2 \text{ s}^{-1}$

Use Torque $= I\dfrac{d^2\theta}{dt^2} = I\alpha$ where $\alpha = \dfrac{d^2\theta}{dt^2}$

But $\omega^2 = \omega_0^2 + 2\alpha\theta$

where $\omega_0 = $ initial ang. velocity $= 20\pi r \text{ s}^{-1}$

$\omega = $ final ang. velocity $= 0 \text{ r s}^{-1}$

$\alpha = $ angular acceleration $-$ to be found

$\theta = $ angular displacement $= 50 \times 2\pi = 100\pi r$

Hence $0 = 400\pi^2 + 2\alpha \times 100\pi$

$\therefore \quad \alpha = -\dfrac{400\pi^2}{200\pi} = 2\pi r \text{ s}^{-2}$

(The minus sign indicating an angular retardation.)

$\therefore$ Torque $= 0.2 \times 2\pi = 0.4\pi \text{ N m}$

Answers: 394·8 J; 4π kg m^2 s^{-1}; 0·4 N m.

3 A particle describes, with period T, a circular orbit which is very near to the moon, and is in a plane containing its centre. If the moon has a radius R and mean density ρ, show that T is a function of ρ but not of R. Calculate, in terms of R and T, the minimum velocity with which the particle should be projected from the moon in order to escape from its gravitational field.

Let $M_m = $ mass of moon

$M_p = $ mass of particle

$\omega = $ angular velocity of particle

$F = $ force acting between moon and particle

We have $F = \dfrac{G M_m M_p}{R^2}$ Newton's Law

But $F = M_s R \omega^2$ Circular Motion

For equilibrium

$$\omega^2 M_s R = \dfrac{G M_m M_s}{R^2}$$

or $\omega^2 = M_m \dfrac{G}{R^3}$

But $M_m = \dfrac{4}{3}\pi R^3 \rho$

$$\therefore \omega^2 = \frac{4\pi R^3 \rho G}{3R^3} = \frac{4\pi \rho G}{3}$$

or
$$\omega = \sqrt{\frac{4\pi \rho G}{3}}$$

Now
$$T = \frac{2\pi}{\omega} = 2\pi \sqrt{\frac{3}{4\pi \rho G}} = \sqrt{\frac{3\pi}{\rho G}} = k\rho^{1/2}$$

where
$$k = \sqrt{\frac{3\pi}{G}}$$

Thus T is $f(\rho)$ but not $f(R)$.

Now the escape velocity v is given by:

$$v = \sqrt{2gR}$$

We need to substitute for g in terms of T

Now
$$g = \frac{G M_m}{R^2} = \frac{4\pi R^3 \rho G}{3R^2}$$
$$= \frac{4\pi R \rho G}{3}$$

We have shown that:

$$\omega = \sqrt{\frac{4\pi \rho G}{3}} \quad \text{i.e.} \quad \omega^2 = \frac{4\pi \rho G}{3}$$

or
$$4\pi \rho G = 3\omega^2$$

$$\therefore g = \frac{3\omega^2 R}{3} = \omega^2 R$$

And
$$T = \frac{2\pi}{\omega} \quad \therefore \omega = \frac{2\pi}{T} \quad \text{or} \quad \omega^2 = \frac{4\pi^2}{T^2}$$

hence
$$g = \frac{4\pi^2 R}{T^2}$$

So:
$$v = \sqrt{2 \frac{4\pi^2 R}{T^2} R}$$
$$= \sqrt{\frac{8\pi^2 R^2}{T^2}}$$

$$= \sqrt{8}\,\frac{\pi R}{T}$$

or

$$v = \sqrt{8\pi^2}\,\frac{R}{T}\,\mathrm{m\,s^{-1}}$$

4   Two identical horizontal circular discs of mass 2 kg each are held at rest on a rough horizontal table with their nearest points 0·3 m apart. These points are connected by a thin cord of natural length 0·1 m having an elastic constant $2\cdot5 \times 10^2\,\mathrm{N\,m^{-1}}$. If the discs are simultaneously released determine the velocity of each at the moment of impact with the other assuming that the cord has not exceeded its elastic limit and the coefficient of kinetic friction between the discs and the table is 0·45 (g = 10 m s^{-2}).

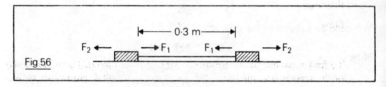

Fig 56

The situation is shown in Fig. 56.

The physics of the problem is that each mass is subjected to an accelerating force supplied by the stretched cord and a retarding force caused by friction between it and the table. The resultant force accelerates each mass.

To find F, the accelerating force

$$F_1 = kx \qquad \text{Hooke's Law}$$
$$k = \text{elastic constant}, \quad x = \text{extension}$$
$$k = 2\cdot5 \times 10^2\,\mathrm{N\,m^{-1}}$$

given x = 0·2 m, since the extension is 0·3 m − 0·1 m

∴ $F_1 = 2\cdot5 \times 10^2 \times 0\cdot2$

$F_1 = 50\,N$

To find F_2, the frictional retarding force

$$F_2 = \mu R$$

where μ = Coefficient of kinetic friction
R = normal reaction
= 0·45 × 2 × 10

∴ $F_2 = 9\,N$

Resultant force, $F_1 - F_2$, acting on each body

$$= 50 - 8 = 41 \text{ N}$$

The acceleration is given by

$$F = m\,a$$

where F = force,   m = mass,   a = acceleration

$$\therefore \ a = \frac{F}{m} = \frac{41}{2} = 20 \cdot 5 \text{ m s}^{-2}$$

We calculate the impact velocity using

$$v^2 = u^2 + 2\,a\,s$$

where v = final velocity,   u = initial velocity,   a = acceleration, s = distance

Now
$$u = 0 \text{ m s}^{-1}$$

$$a = 20 \cdot 5 \text{ m s}^{-2}$$

To find s, we use the fact that the extension of the cord is $0 \cdot 2$ m. Since both masses are identical they will each travel identical distances before impact.

$$\therefore \ s = 0 \cdot 1 \text{ m}$$

Hence
$$v^2 = 2 \times 20 \cdot 5 \times 0 \cdot 1$$

$$v = 4 \cdot 1$$

$$v = 2 \cdot 025 \text{ m s}^{-1}$$

Velocity at impact = $2 \cdot 025 \text{ m s}^{-1}$

5   A falling sphere experiment is performed twice – firstly using olive oil and secondly using castor oil with the same steel sphere used throughout. Calculate the ratio of the spheres terminal velocity in olive oil to that in castor oil.

Assume density of steel = 7800 kg m^{-3} = ρ
  density of olive oil = 920 kg m^{-3} = δ_1
  density of castor oil = 940 kg m^{-3} = δ_2
  coeff. of viscosity of olive oil = $8 \cdot 4 \times 10^{-2}$ N m^{-2} s = η_1
  coeff. of viscosity of castor oil = $2 \cdot 42$ N m^{-2} s = η_2

Use

$$v = \frac{2a^2 g}{9\eta}(\rho - \delta)$$

If v_1 = terminal velocity in olive oil
        v_2 = terminal velocity in castor oil

then $v_1 = \dfrac{2a^2g}{9\eta_1}(\rho - \delta_1)$

        $v_2 = \dfrac{2a^2g}{9\eta_2}(\rho - \delta_2)$

$\therefore \quad \dfrac{v_1}{v_2} = \dfrac{\eta_2}{\eta_1}\dfrac{(\rho - \delta_1)}{(\rho - \delta_2)}$

Substituting the values

$\dfrac{v_1}{v_2} = \dfrac{2 \cdot 42}{8 \cdot 4 \times 10^{-2}} \times \dfrac{6880}{6860}$

$= 28 \cdot 89$

$\therefore$ **Ratio of velocities** $= 28 \cdot 89 : 1$

6   If the radius of a disc is 3×10^{-2} m and its mass is 9×10^{-3} kg, calculate its total energy when it rolls along a table with a velocity of 6 $\times 10^{-2}$ m s^{-1}.

Use

Total K.E. $= \dfrac{1}{2}I\omega^2 + \dfrac{1}{2}Mv^2$

        I = Moment of Inertia
        ω = Angular velocity
        M = mass
        v = translational velocity
        a = radius

Now $I = \dfrac{Ma^2}{2} = \dfrac{9 \times 10^{-3} \times (3 \times 10^{-2})^2}{2}$

        $= \dfrac{81 \times 10^{-7}}{2} = 4 \cdot 05 \times 10^{-6} \text{ kg m}^2$

        $\omega = \dfrac{v}{r} = \dfrac{6 \times 10^{-2}}{3 \times 10^{-2}} = 2 \text{ r s}^{-1}$

$\therefore$ Total Energy

$= \left[\dfrac{1}{2} \times 4 \cdot 05 \times 10^{-6} \times 4\right] + \left[\dfrac{1}{2} \times 9 \times 10^{-3} \times (6 \times 10^{-2})^2\right]$

$$= 2 \times 4 \cdot 05 \times 10^{-6} + 9 \times 18 \times 10^{-7}$$
$$= 8 \cdot 1 \times 10^{-6} + 16 \cdot 2 \times 10^{-6}$$
$$= 24 \cdot 3 \times 10^{-6} \, \text{J}$$
Total Energy $= \mathbf{24 \cdot 3 \times 10^{-6} \, J}$

7 A mass of $0 \cdot 3$ kg is constrained to move in a vertical circle by a steel wire of area of cross section 10^{-6} m^2 and length 1 m. If the wire just remains taut throughout the motion, calculate (i) the value of the velocity of the mass at the highest (ii) lowest points of its path and (iii) the maximum increase in length of the wire during the rotation. Assume Young's Modulus for steel $= 2 \times 10^{11}$ Pa., $g = 10$ m s^{-2}. The situation is shown in Fig. 57

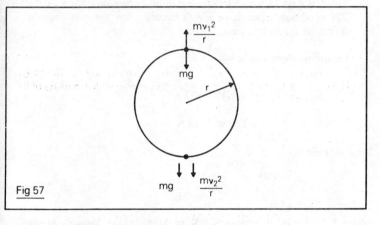

Fig 57

Let v_1 = velocity at highest point
v_2 = velocity at lowest point
If the wire just remains taut, then at the highest point
Upward force = Downward force
i.e. Centrifugal force = Weight

$$\frac{mv_1^2}{r} = m g$$

$$v_1 = \sqrt{g r}$$

Substituting the values

$$v_1 = \sqrt{10 \times 1} = \sqrt{10} = 3 \cdot 162 \, \text{m s}^{-1}$$

(i) Velocity at highest point $= 3\cdot162\,\mathrm{m\,s^{-1}}$

Now as the mass moves from the highest to the lowest point, potential energy is converted into extra kinetic energy.

i.e.
$$\frac{1}{2}m\,(v_2^{\,2} - v_1^{\,2}) = m\,g\,h$$

$$v_2^{\,2} = v_1^{\,2} + 2\,g\,h$$

$$h = 2\,\mathrm{m}$$

$\therefore$
$$v_2^{\,2} = 10 + 2 \times 10 \times 2 = 50$$

$\therefore$
$$v_1 = \sqrt{50} = 7\cdot071\,\mathrm{m\,s^{-1}}$$

(ii) Velocity at lowest point $= 7\cdot071\,\mathrm{m\,s^{-1}}$

Note: We could have used the equation of motion $v^2 = u^2 + 2\,a\,s$. The maximum increase in length occurs when the maximum force occurs, at the lowest point.

$$\text{Maximum force} = m\,g + \frac{mv_2^{\,2}}{r}$$

$$= 0\cdot3 \times 10 + 0\cdot3 \times \frac{50}{1}$$

$$= 0\cdot3 \times 60 = 18\,\mathrm{N}$$

Now extension,

$$e = \frac{Fl}{AE}$$

where $F = $ force, $l = $ length, $A = $ Area, $G = $ Young's Modulus

Substituting values

$$e = \frac{18 \times l}{10^{-6} \times 2 \times 10^{11}} = 9 \times 10^{-5}\,\mathrm{m}$$

(iii) Maximum extension $= 9 \times 10^{-5}\,\mathrm{m}$

8   A particle is dropped down a deep shaft which extends to the centre of the earth. Find its velocity when it reaches the bottom. Assume radius of the earth $= 6\cdot4 \times 10^6\,\mathrm{m}$, acceleration due to gravity on the surface of the earth $= 10\,\mathrm{m\,s^{-2}}$.

Inside the earth:

Acceleration due to gravity α Distance from centre of earth

Further, it acts to reduce the distance

∴ Acceleration = −constant × displacement

i.e. the motion is S.H.M., of amplitude equal to the radius of the earth and equilibrium point corresponding to the centre of the earth.

The maximum velocity will be at the centre of the earth.

At the centre of the earth

$$V_{max} = W a$$

Now

$$a = 6·4 \times 10^6 \text{ m}$$

In order to find W we use:

$$W = \sqrt{\frac{\text{acceleration}}{\text{displacement}}}$$

$$= \sqrt{\frac{10}{6·4 \times 10^6}}$$

Hence:

$$v_{max} = \sqrt{\frac{10}{6·4 \times 10^6}} \, 6·4 \times 10^6$$

$$= \sqrt{6·4 \times 10^7}$$

$$= 8 \times 10^3 \text{ m s}^{-1}$$

so, Velocity at centre of earth $= 8 \times 10^3$ m s^{-1}

9 A mass of 1 kg suspended by a cord 1 m long is made to describe a horizontal circle four times every second. Calculate the extension of the cord, given Young's Modulus for rubber is 10^7 Pa., Area of cord $= 5 \times 10^{-5}$ m^2.

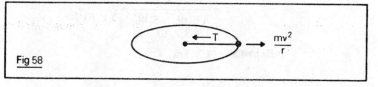

Fig 58

In Fig. 58 we have the tension in the cable, T is given by

$$T = \frac{mv^2}{r}$$

m = mass, v = velocity, r = radius, F = Force, l = length, A = Area, e = extension

Using
$$E = \frac{Fl}{Ae}$$

Hence
$$T = F = \frac{AeF}{l}$$

so
$$\frac{AeE}{l} = \frac{mv^2}{r}$$

$\therefore$
$$e = \frac{mv^2l}{AEr} = \frac{mv^2}{AE} \quad \text{since } l = r$$

Now
$$v = r\omega \quad (\omega = \text{angular velocity})$$

$$v^2 = r^2\omega^2$$

$\therefore$
$$e = \frac{mr^2\omega^2}{AE}$$

Substituting
$$m = 1 \text{ kg}$$
$$r = 1 \text{ m}$$
$$\omega = 4 \text{ revs s}^{-1} = 8\pi r \text{ s}^{-1}$$
$$A = 5 \times 10^{-5} \text{ m}^2$$
$$E = 10^7 \text{ Pa}$$
$$e = \frac{1 \times 1 \times 64\pi^2}{5 \times 10^{-5} \times 10^7} = \frac{64\pi^2}{5 \times 10^2}$$
$$= 126 \cdot 3 \times 10^{-2} \text{ m}$$

Extension = 1·263 m

10 A U tube which has its ends open and its limbs vertical, contains a liquid of surface tension $2 \cdot 5 \times 10^{-2} \text{ N m}^{-1}$ and density 800 kg m^{-3}, the angle of contact being zero.

The radius of one limb is $2 \times 10^{-4} \text{ m}$ and the other is 10^{-4} m. Find the difference in levels of the liquid in the limbs. Take $g = 10 \text{ m s}^{-2}$.

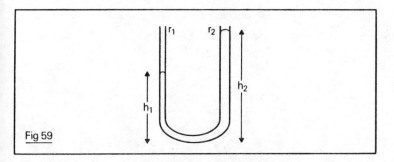

Fig 59

r_1 = radius of narrow limb

r_2 = radius of wide limb

h_1 = height in narrow limb

h_2 = height in wide limb

T = surface tension of liquid

Pressure at bottom of limb = Hydrostatic Pressure + Excess
Pressure due to surface tension

At equilibrium, the pressures at the bottom of each limb are identical.

$$\therefore \; \rho g h_1 + \frac{2T}{r_1} = \rho g h_2 + \frac{2T}{r_2}$$

because Cos $\theta = 1$ since the angle of contact is zero

$$\therefore \; 2T\left(\frac{1}{r_1} - \frac{1}{r_2}\right) = \rho g (h_2 - h_1)$$

$$\therefore h_2 - h_1 = \frac{2T}{\rho g}\left(\frac{1}{r_1} - \frac{1}{r_2}\right)$$

Substituting the values

$$h_2 - h_1 = \frac{2 \times 2 \cdot 5 \times 10^{-2}}{800 \times 10}\left(\frac{1}{10^{-4}} - \frac{1}{2 \times 10^{-4}}\right)$$

$$= \frac{5 \times 10^{-2}}{8 \times 10^3} \times 5 \times 10^3$$

$$= 3 \cdot 125 \times 10^{-2}\,\text{m}$$

But $h_2 - h_1$ = difference in height

∴ **Difference in height** $= 3 \cdot 125 \times 10^{-2}$ **m**

13 OMNIBUS PRACTICE QUESTIONS

1 A body of mass 2×10^{-2} kg is attracted towards a fixed point with a force equal to 0.3 N when it is 5×10^{-2} m away from the point. If the motion is simple harmonic, find its period. If the amplitude is 10^{-1} m, determine the maximum kinetic energy of the mass.
Assume $g = 10$ m s^{-2}

2 If the mean density of the moon is 0.6 times the mean density of the earth and the mean radius of the earth is $11/3$ times the mean radius of the moon, calculate a value for the length of a simple pendulum, on the moon, which has a period of 1 s.

3 Calculate the pressure inside an air bubble of radius 4×10^{-4} m at a depth of 12×10^{-2} m in a vessel of water.
Surface tension of water $= 7.26 \times 10^{-2}$ N m^{-1}
Atmospheric pressure $= 0.76$ m Hg
Density of mercury $= 13,600$ kg m^{-3}
Density of water $= 1000$ kg m^{-3}
$g = 9.81$ m s^{-2}

4 The earth's equatorial radius is about 21 km greater than its polar radius, a phenomenon known as the 'Equatorial Bulge'. If the earth were to rotate twice as fast as it does now, would you expect the bulge to be more than, the same as or less than it is now? Explain.

1 1 $M L^2 T^{-2}$

   2 No dimensions. It is a ratio.

   3 Show your working for this question

   4 $v = k\sqrt{g\lambda}$, v = velocity, g = acc. grav., λ = wavelength

2 1 $2 \, m \, s^{-1}$

   6 $4 \, m \, s^{-2}$

   3 $12 \, m \, s^{-1}$, $0.4 \, m \, s^{-2}$

   4 $3 \, s$; $5 \, m$

3 1 $14 \, km$

   2 17.1 units; $95°\ 16'$

   3 $8.66 \, m \, s^{-1}$; $11.55 \, m \, s^{-1}$

4 1 $1 \, m \, s^{-2}$

   2 $10^{-2} \, kg$

   3 $1.5 \, m \, s^{-1}$ in direction of larger mass

5 1 $1 \, kw$

   2 $9.822 \, m \, s^{-1}$

6 1 Resultant couple is 0, hence no rotation

   2 $52 \, N$

   3 $1.699 \, m \, s^{-1}$; $0.2309 \, N$

   4 $4 \, J$

   5 $7.648 \times 10^{-3} \, kg \, m^2 \, s^{-1}$

   6 $16:1$

   7 $20 \, r \, s^{-2}$

   8 $\dfrac{1}{\sqrt{3}} \, (m)$

   9 $\dfrac{7}{5} M a^2 \, (kg \, m^2)$

  10 $\dfrac{5}{7} g \, Sin \, \alpha \, (m \, s^{-2})$

7 1 $1.405 \, s$

   2 $0.32\pi \, m \, s^{-1}$; $64\pi \, m \, s^{-2}$

   3 $4 \times 10^{-2} \, m$

   4 $1.25 \times 10^{-3} \, J$

   5 $1.25 \, m$

8 1 $0.335 \, m$

   2 $272 \, m \, s^{-2}$

   3 $-1.26 \times 10^6 \, J$; $1.59 \times 10^3 \, m \, s^{-1}$

9 1 $10 \, N$

   2 The same

   3 $8.482 \times 10^{-2} \, N$

4   $25\cdot7$ s

10 1 $1\cdot05146 \times 10^5$ N

2   $1\cdot069 \times 10^{-4}$ m

3   $8\cdot4 \times 10^{-2}$ N m^{-1}

11 1 $0\cdot2828\pi$ s

2   $1\cdot273 \times 10^7$ N; $1\cdot2 \times 10^{-4}$; $1\cdot061 \times 10^{11}$ Pa

3   $1\cdot3 \times 10^9$ Pa

4   $405\cdot3$ J

5   $4\cdot4 \times 10^7$ Pa

13 1 $0\cdot363$ s; 3×10^{-2} J

2   $4\cdot155 \times 10^{-2}$ m

3   $1\cdot029362 \times 10^{-5}$ N

4   The bulge would be larger owing to the greater centrifugal force.

IMPROVE YOUR
W
CELTIC REVI

In a tough world, every qualification counts towards a brighter future. So naturally, you want to do all you can to get through those exams.

And now, with *Celtic Revision Aids*, you can really do something about it.

By choosing titles from those series in the Celtic range which are appropriate to you, you can plan long-term course preparation, or carry out last minute revision. You can study model answers and essay guidelines, or check your knowledge against lists of basic facts.

Celtic Revision Aids are a complete range of books, covering subjects both in depth and on an 'essential fact' level.

In short, they're designed to make revision easier, designed to help you to examination success.

Series and titles available:

Rapid Revision Notes O-level

£1.50 each

Titles: *English Language Book 1, English Language Book 2, Mathematics, Physics, Chemistry, Biology, Human Biology, Physical Geography, Commerce, Economics, Sociology, British Economic History, Integrated Science, Commercial Mathematics, Accounts.*

Rapid Revision Notes A-level

£1.75 each

Titles: *Pure Mathematics, Applied Mathematics, Statistics, General Biology, Botany, Zoology, Inorganic Chemistry, Organic Chemistry, Physical Chemistry, Physics – Mechanics, Physics – Heat, Light and Sound, Physics – Electricity and Magnetism.*

EXAM RESULTS
H
SION AIDS

Literature Revision Notes and Examples
£1.50 each
Titles: *Chaucer's Prologue, Merchant of Venice, Julius Caesar, Richard II, A Midsummer Night's Dream.*

New Testament Studies
£1.50 each
Titles: *St. Matthew, St. Mark, St. Luke, St. John, Acts of the Apostles.*

Law Revision Notes
£1.95
Titles: *Principles of Law, Criminal Law, Family Law, Law of Tort, Company Law.*

Model Answers
£1.25 each
Titles: *Julius Caesar, Macbeth, Romeo & Juliet, Merchant of Venice, Aids to Mathematics, Aids to Essay Writing, English Language, Model Essays, Practice in Summary Writing, Essay Plans, Biology, Chemistry, Human Biology, Physics, Mathematics, Commerce, Economics, British Isles Geography, British Economic History, Accounts, Commercial Mathematics, Integrated Science.*

Test Yourself
95p each
Titles: *English Language 1, English Language 2, French, German, Commerce, Economics, Chemistry, Physics, Mathematics, Modern Mathematics, Biology, Human Biology, St. Matthew, St. Mark, St. Luke, St. John, Acts of the Apostles, Commercial Mathematics, Accounts, Statistics, British Isles Geography, British Economic History.*

Multiple Choice O-level

£1.25 each

Titles: *English 1, English 2, French, Mathematics, Modern Mathematics, Chemistry, Physics, Biology, Human Biology, Commerce, Economics, British Isles Geography, Accounts, Commercial Mathematics, Integrated Science.*

Multiple Choice A-level

£1.50 each

Titles: *Pure Mathematics, Applied Mathematics, Chemistry, Physics, Biology, Statistics.*

Worked Examples A-level

£1.50 each

Titles: *Pure Mathematics, Applied Mathematics, Chemistry, Physics, Biology, Economics 1, Economics 2, Sociology, British History 1914-76, European History 1914-76, British Economic History, Physical Geography, Accounts, Statistics.*

For a full colour brochure giving details of the complete range of Celtic Revision Aids, send your name and address to:

Celtic Revision Aids – Direct Brochure Mailing, Dept. T.,
TBL Book Service Ltd.,
17-23 Nelson Way,
Tuscam Trading Estate,
Camberley,
Surrey.
GU15 3EU

Celtic Revision Aids are available from good booksellers everywhere or, in cases of difficulty in obtaining them, direct from the publisher. If you wish to order books direct from Celtic, write to the above address giving your own name and address in block capitals, clearly stating the title/s of the book/s you would like, and enclosing a cheque or postal order made payable to TBL Book Service Ltd., to the value of the cover price of the books required *plus* 25p postage and packing per order for orders up to 4 books. (Postage and packing is free for orders of 5 books or more.) (U.K. only.)

Celtic Revision Aids reserve the right to show new retail prices on covers which may differ from those previously advertised in the text or elsewhere, and to increase postal rates in accordance with the P.O.